The Human Machine:
A Systems Perspective

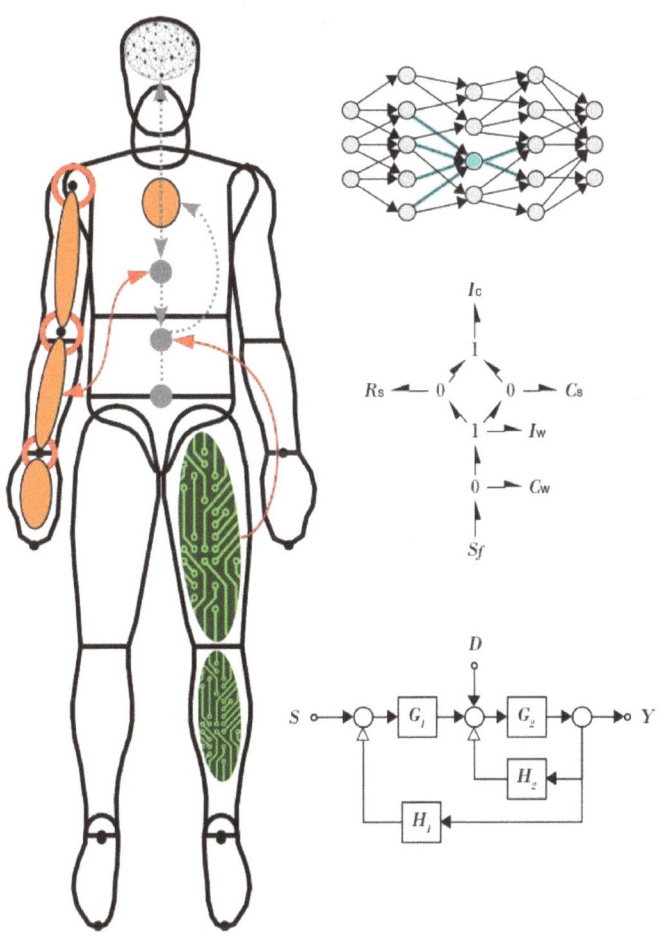

David R. Monteverde

The Human Machine: A Systems Perspective
Copyright ©2021 by David R. Monteverde

davro@alum.mit.edu

May 2021
San Diego, California

Version 1.0

Lulu Press
Printed in the United States of America

ISBN: 978-1-6671-1397-5

Contents

Bibliography

[Ber69]   Ludwig von Bertalanffy. *General System Theory*. George Braziller Publishers, 1969.

[BG12]   Erich Blechschmidt and R.F. Gasser. *Biokinetics and Biodynamics of Human Differentiation*. North Atlantic Books, 2012.

[BHU11]   Florian Beissner, Christian Henke, and Paul Unschuld. "Forgotten Features of Head Zones and Their Relation to Diagnostically Relevant Acupuncture Points". In: *Evidence-Based Complementary and Alternative Medicine* 2011:24065 (2011).

[BLP16]   Domenico Bruno, Paolo Lonetti, and Arturo Pascuzzo. "An Optimization Model for the Design of Network Arch Bridges". In: *Computers and Structures* 170 (2016).

[BM14]   Erik Brynjolfsson and Andrew McAfee. *The Second Machine Age*. W. W. Norton & Company, 2014.

[Bro18]   Robert Brooker. *Genetics, Sixth Edition*. McGraw-Hill, 2018.

[CB07]   Dennis Carter and Gary Beaupré. *Skeletal Function and Form*. Cambridge University Press, 2007.

[CO14]   Bradley Carroll and Dale Ostlie. *An Introduction to Modern Astrophysics*. Pearson, 2014.

[CUK02]   Eric Cohen, Kamil Ugurbil, and Seong-Gi Kim. "Effect of Basal Conditions on the Magnitude and Dynamics of the Blood Oxygenation Level–Dependent fMRI Response". In: *Journal of Cerebral Blood Flow & Metabolism* 22 (2002).

[Dav19]   Paul Davies. *The Demon in the Machine*. The University of Chicago Press, 2019.

[Dim11]   Theodore Dimon. *The Body in Motion*. North Atlantic Books, 2011.

[DP11]   Vanessa Díaz-Zuccarini and César Pichardo-Almarza. "On the Formalization of Multi-scale and Multi-science Processes for Integrative Biology". In: *Interface Focus* 1 (2011).

[GCC15]   Peter Gawthrop, Joseph Cursons, and Edmund Crampin. "Hierarchical Bond Graph Modelling of Biochemical Networks". In: *Proceedings of the Royal Society A* 471:20150642 (2015).

[GG17]   Umut Güçlü and Marcel van Gerven. "Modeling the Dynamics of Human Brain Activity with Recurrent Neural Networks". In: *Frontiers in Computational Neuroscience* 11 (2017).

[GK17]   David Goldstein and Irwin Kopin. "Homeostatic Systems, Biocybernetics, and Autonomic Neuroscience". In: *Autonomic Neuroscience* 208:15-28 (2017).

[Gre07]   Brian Greene. *The Fabric of the Cosmos*. Random House, 2007.

[Hea97]   National Institutes of Health. *Acupuncture: NIH Consensus Statement*. Tech. rep. U.S. Department of Health & Human Services, 1997.

[II10]   Pablo Iglesias and Brian Ingalls, eds. *Control Theory and Systems Biology*. The MIT Press, 2010.

[Kar15]   Kenneth Kardong. *Vertebrates: Comparative Anatomy, Function, Evolution*. McGraw-Hill, 2015.

[Keo14]   Daniel Keown. *The Spark in the Machine*. Kingsley Publishers, 2014.

[KMR12]   Dean Karnopp, Donald Margolis, and Ronald Rosenberg. *System Dynamics: Modeling, Simulation, and Control of Mechatronic Systems*. Wiley, 2012.

[KO12]   Seong-Gi Kim and Seiji Ogawa. "Biophysical and Physiological Origins of Blood Oxygenation Level-Dependent fMRI Signals". In: *Journal of Cerebral Blood Flow & Metabolism* 32, 1188–1206 (2012).

[KT+13]   W. Kean, S. Tocchio, et al. "The Musculoskeletal Abnormalities of the Similaun Iceman (Ötzi): Clues to Chronic Pain and Possible Treatments". In: *Inflammopharmacology* 21 (2013).

[Lin+09]   Martin Lindquist et al. "Modeling the Hemodynamic Response Function in fMRI: Efficiency, Bias and Mis-modeling". In: *NeuroImage* 45 (2009).

[MA14]   Johnjoe McFadden and Jim Al-Khalili. *Life on the Edge: The Coming of Age of Quantum Biology*. Crown Publishing, 2014.

[Mit08]   Edgar Mitchell. *The Way of the Explorer*. New Page Books, 2008.

[MK17]   Alison Marsden and Ethan Kung. "Multi-scale Modeling of Cardiovascular Flows". In: *Computational Bioengineering* (2017).

[Mon12]   David Monteverde. *Kinetics for Robotics and Biomechanics*. Lulu Press, 2012.

[MWK16]   Adam Marblestone, Greg Wayne, and Konrad Kording. "Toward an Integration of Deep Learning and Neuroscience". In: *Frontiers in Computational Neuroscience* 10 (2016).

[Oga10]   Katsuhiko Ogata. *Modern Control Engineering*. Prentice Hall, 2010.

[Org03]   World Health Organization. *Acupuncture: Review and Analysis of Reports on Controlled Clinical Trials*. Tech. rep. WHO Library, 2003.

[Pen+17]   Roger Penrose et al. *Consciousness and the Universe: Quantum Physics, Evolution, Brain & Mind*. Cosmology Science Publishers, 2017.

[Pla68]   Plato. *The Republic*. Trans. by Alan Bloom. Harper Collins, 1968.

[Pur+08]   Dale Purves et al. *Neuroscience, Fourth Edition*. Sinauer, 2008.

[Raa20]   Panu Raatikainen. "Gödel's Incompleteness Theorems". In: *The Stanford Encyclopedia of Philosophy* (2020). Ed. by Edward N. Zalta.

[Sch+15] Gary Schoenwolf et al. *Larsen's Human Embryology*. Elsevier, 2015.

[Sch44] Erwin Schrödinger. *What is Life?* Cambridge University Press, 1944.

[She15] Scott M. Shell. *Thermodynamics and Statistical Mechanics: An Integrated Approach*. Cambridge University Press, 2015.

[Sto19] James Stone. *Artificial Intelligence Engines*. Sebtel Press, 2019.

[Su+09] B. Su et al. "Mathematical Modelling of Immune Response in Tissues". In: *Computational and Mathematical Methods in Medicine* 10 (2009).

[VB13] Alejandro Villaverde and Julio Banga. "Reverse Engineering and Identification in Systems Biology: Strategies, Perspectives and Challenges". In: *J R Soc Interface* 11 (2013).

[Voi20] Eberhard Voit. *Systems Biology: A Very Short Introduction*. Oxford University Press, 2020.

[Wan10] Ingrid Wankura-Kampik. *Segmental Anatomy*. Elsevier Urban & Fischer Verlag, 2010.

[Wie65] Norbert Wiener. *Cybernetics: Control and Communication in the Animal and the Machine*. The MIT Press, 1965.

[Wol02] Stephen Wolfram. *A New Kind of Science*. Wolfram Media, 2002.

[WPK09] Charles Watson, George Paxinos, and Gulgun Kayalioglu. *The Spinal Cord*. Academic Press, 2009.

[ZB14] Wei Zhou and Peyman Benharash. "Effects and Mechanisms of Acupuncture Based on the Principle of Meridians". In: *J Acupunct Meridian Stud* 7(4):190e193 (2014).

Introduction

How and why machines work has been a fascination of mine from a very early age. And being the operator of a particularly amazing machine –the human body– my curiosity extended to living organisms. That interest in descriptive knowledge was combined with a passion to build and create, so eventually I was drawn into engineering.

This led me to pursue studies in the areas of robotics and biomechanics at MIT and later at Stanford University, which opened my eyes to various cross-functional fields. Additionally, during the course of my career, I had the opportunity to develop spacecraft control systems, provide biomechanical consulting services, and design surgical robots.

A couple of truly valuable lessons that resulted from these experiences was how to look at problems from contrasting perspectives, and how to leverage tools from very diverse sources to overcome difficult obstacles. This has given me a particular worldview, that I like to think of as having a certain *cosmopolitan* flavor.

A previous book that I wrote on the general topic of "the human body as machine" employed an utilitarian approach of leveraging the well-developed methods of spacecraft design and analysis, and applying them to robotics and biomechanics.[1] By its nature, it is detailed in the mathematics and practical uses.

The present book is a second synthesis, much broader in perspective, and much more conceptual than mathematical. I hope that the story is interesting and enjoyable, and that some of the observations and ideas presented here can help the reader gain a worldview that is *just a bit different*.

---

[1][Mon12]

Acknowledgements

I would like to express my gratitude to Matt Williams, Elias Polendo, Mark Durst and Akiko Kurihara for reviewing the draft manuscript and providing valuable feedback.

Part I

Systemic Foundations of Life

Chapter 1

Body as a System

The human body is a machine. This machine is composed of organs, which in turn are made of tissues, cells, molecules, atoms, and subatomic particles. And like any physical entity, the human body depends on physical laws.

The laws of physics are our basic tool for understanding the functioning of any machine, including biological organisms. Biology is, after all, applied chemistry; and chemistry is really just applied physics.[1]

But physical laws by themselves are not sufficient to characterize physical mechanisms. Interaction between elements is of fundamental importance, and a machine is generally better described as a system. The complex machine that we know as the human body is perhaps optimally studied under the umbrella of *systems theory*.

1.1 Systems Theory

A system can be simply defined as a set of elements in interaction.[2] Systems theory is the interdisciplinary study of such entities.

The theory places particular emphasis on the relations between elements. It also leverages isomorphic (homologous) behaviors in systems of different types, in order to understand complicated and poorly-explained phenomena.

---

[1][MA14, p. 19]
[2][Ber69, p. 83]

Important concepts employed in systems theory include:

- **System:** An entity made of elements in interaction.
- **Boundaries:** Spatial or dynamic borders that define a system.
- **Topology:** Structural arrangement between elements.
- **Hierarchical order:** Arrangement of elements into levels, chosen with respect to some aspect (property) of interest.
- **Isomorphism:** Structural or phenomenological similarities between systems or entities that are, intrinsically, very different.
- **Emergent properties:** Properties exhibited by a system as a whole, but not by any of its individual elements. These properties or behaviors emerge when elements interact.
- **Homeostasis:** The tendency of a system to be resilient with respect to external disruption, and to maintain its key characteristics.
- **Information persistence:** Generalized memory, ranging from short-term computer memory such as RAM to phylogenetic learning enabled by DNA.
- **Amplification:** Mechanism by which a small change in an element causes large changes in the system.
- **Inhibition:** Mechanism by which a process is restricted or obstructed within a system. Generally, an opposite effect to amplification.
- **Feedback Loop:** Process by which a system self-corrects, based on circular interaction between elements.

Systems theory has been successfully applied in engineering, biology, economics, and even psychology. But from among these, perhaps the most successful application has been in engineering sciences, where precise mathematical formulations have enabled widespread use. Some of these include:

- **Systems engineering:** Discipline used to guide the development of highly complex machinery, ranging from medical devices to spacecraft. These systems are developed in the confluence of heterogeneous technologies, and systems engineering leverages expertise from diverse fields such as mechatronics, information technology, chemical engineering, medicine, and others.

- **Control theory:** Field of study that deals with control actions intended to shape the dynamics of a system. In closed-loop control schemes, deviation from desired states (errors) are used in feedback algorithms to regulate system behavior.[3]

- **Bond graph:** Analytical tool that models power flow dynamics in multi-domain systems. Representation of variables is domain-independent, so it can easily handle the modeling of systems containing elements that are mechanical, electrical, chemical, etc. in character. For instance, the generalized quantity of *effort* is a universal abstraction for force (mechanical domain), pressure (hydraulic domain), voltage (electrical domain) and chemical potential (chemical domain). Bond graphs are used to produce sets of coupled differential equations, which can be further abstracted in matrix equations and studied via computer simulations.[4]

- **Biomechanics:** The study of biological systems using analytical methods, primarily from mechanical engineering. It draws strongly on isomorphisms to map well-developed engineering techniques onto biological entities. For instance, biomechanics studies the motion of multi-articulated animals, the structural integrity of bone, and the hydrodynamics of blood flow.[5]

Systems thinking inspired developments that have made important aspects of our modern life possible. And many more advancements may still lie ahead.

Over the last century or so, scientific and technical knowledge has incremented by staggering amounts. This has led to increased depth in various branches of knowledge, along with growing numbers of highly-specialized individuals. And our pace of development has been so rapid that we've had to augment human specialized knowledge with artificial computing power, to the point where computers are now regularly performing non-trivial intellectual tasks.

Just as the industrial revolution resulted in the replacement of much physical human effort by high-power machinery, it looks like the replacement of most routine intellectual tasks by high-power computation is the next evolutionary step in our culture. We are currently transitioning into

---

[3][Oga10]
[4][KMR12], [GCC15]
[5][Mon12], [CB07], [MK17]

what has been called the *second machine age*.[6]

Access to abundant and advanced computing power in the second machine age opens the door to interesting new possibilities. Modeling and simulations that were not feasible or practical previously could help us gain new insights and allow us to understand phenomena that escaped us before.

And this is the sweet spot for systems theory: taking on areas unexplored and neglected as "no man's land" between the various established fields.[7] To illustrate the potential applicability of systems thinking towards the development of novel branches of knowledge, we will use acupuncture as a case study (in Part II of the book). Acupuncture is interesting because of its objective effectiveness, which stands in contrast to the vagueness and lack of consensus in terms of credible mechanisms and models to explain it. As we will see later, acupuncture has clear phenomenological isomorphisms with closed-loop control schemes, which puts it squarely in the purview of cybernetics (a branch of systems theory that studies self-regulating mechanisms in animals).

Humans are storytellers by nature, so it is pertinent that the earlier assertion that "the human body is a machine" should be expanded in the form of a story (apart from the fact that I like telling stories). And the story starts by paraphrasing American systems theorist Buckminster Fuller: *to understand the human machine, one must begin with the universe*.[8]

---

[6][BM14]

[7][Wie65, Intro], [Ber69, p. 99]

[8]Buckminster Fuller reportedly said: "to understand the human condition, one must begin with the universe."[Mit08, p. 47].

Chapter 2

Scales of the Universe

Reality, in the modern conception, appears to be a vast hierarchical order of many levels.[1] Depending on the scale at which we observe, the fabric of the universe can look very different. An illustration of the universe viewed at different scales is shown in Figure 2.1.[2]

The sequence of frames (from large scale to small scale) portrays the following: Milky Way galaxy, inner solar system, planet Earth, city of San Diego, cats, red blood cells, DNA helix, electron cloud, and atomic nucleus.

Concepts and conclusions, including physical laws, that are relevant at one scale may not be fully applicable at other scales.[3] Think of the pixelated image you see when looking at a television screen from a very short distance. Each pixel can change in brightness or in color, but its location doesn't change. However, as you move away from the screen, the pixels start to blur together and you can perceive moving objects, like cars racing or soccer players running around the field. This is a simplistic portrayal, but it encapsulates the basic idea.

The range of scales shown in the figure is overwhelming, to say the least. The magnification factor from the frame on the bottom right to that on the top left is 10^{36}, or 1000000000000000000000000000000000000.

These large changes of scale can be difficult to grasp, so let us use an

---

[1][Ber69, p. 87]

[2]This portrayal of the universe parallels the depiction in the film *Powers of Ten* (Charles and Ray Eames, 1977).

[3][Gre07, p. 334] We will cover this in Section 3.4.

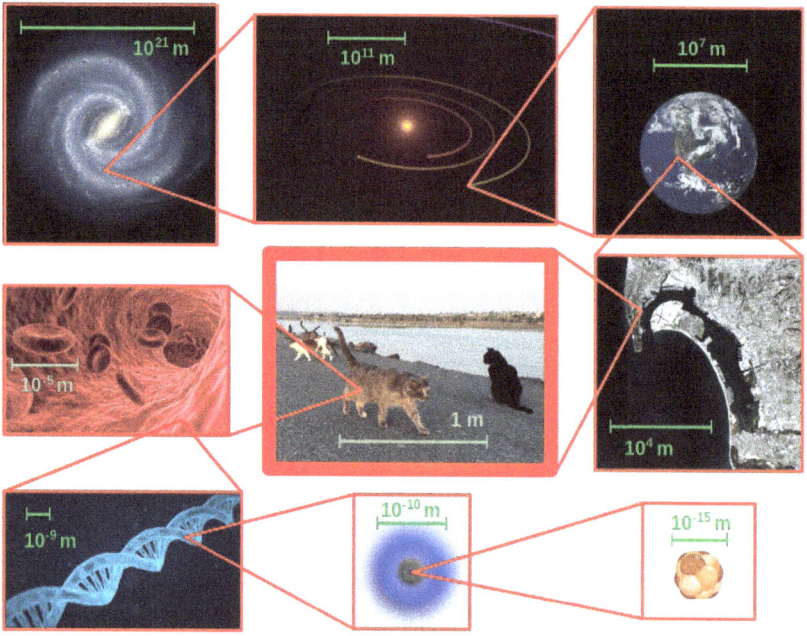

Figure 2.1: Scales of the Universe

analogy to illustrate. First, think of the size of a ballpoint pen tip, which is about 1 mm in diameter (10^{-3} m). Now consider a gigantic sphere that is 10,000 km in diameter (10^7 m), which would make it almost as large as planet Earth (our planet's diameter is about 12,700 km). If you were to magnify the diminutive ballpoint pen tip to the size of the gigantic Earth-sized sphere, you would be performing a scale conversion of 10^{10}, or 10000000000. To traverse all the frames, you would need to magnify that scale by an additional factor of 10^{26}, or 100000000000000000000000000.

And the change of scales keeps going beyond what is shown in Figure 2.1. At scales larger than that of the upper-left frame, you have galaxy clusters, then superclusters, then bubbles and voids.[4]

On the other hand, going towards the very small beyond the lower-right frame, there are subatomic particles (such as quarks, leptons and

---

[4][CO14, pp. 1156, 1167–1178]

bosons), which at present is as far as we can determine experimentally given the current capability of our most powerful instruments. Going further than that, some theories hold that the most elementary constituents of nature may operate down at the scale of Planck length,[5] an ultra-tiny distance of about 1.6×10^{-35} m.

---

[5][Gre07, p. 349]

Chapter 3

Modeling of Nature

Physics is fundamentally concerned with general phenomena that emerge at different levels of reality.[1] As might be obvious from the discussion in the previous chapter, each level is formed by tremendous numbers of particle elements at a lower level, in a recursive fashion. So at any given scale, notions of continuity or smoothness are just idealizations. Yet this kind of idealization is almost required in the formulation of physical laws, otherwise we would not be able to make predictions in a practical way.[2]

Physical laws are human-created models that we use to predict certain natural processes. Some physical laws are called *fundamental laws*, because they are not logically derived from other laws; rather, we accept them due to the fact that these models are capable of making very accurate (and repeatable) predictions of experimental phenomena.[3]

Austrian physicist Erwin Schrödinger (one of the "fathers" of quantum mechanics) observed that physical laws have a statistical character, in many cases due to the interaction of large numbers of particles.[4] He argued that the precision of these laws was based on the number of particles that participated in the observed phenomena. It was because of the large-numbers effect that these models had the accuracy needed to claim the dignity of being called "Laws of Nature".[5]

---

[1] [Sch44, p. 10], [Ber69, p. 113]
[2] [Wol02, p. 729]
[3] [She15, sec. 3.1], [Ber69, p. 83]
[4] He called them "order-from-disorder" laws [Sch44, p. 10]. Also see [MA14, p. 53].
[5] [Sch44, p. 17]

As an example, classical thermodynamics provides a set of laws and relationships that concern bulk properties of a system (such as temperature and pressure of a gas inside a container). To be sure, the bulk properties result from the interactions between many molecules, each of which has individual properties such as position and velocity at any given time. But in using thermodynamics laws, it is the average behavior of lots of molecules that is predictable, not necessarily the behavior of individual molecules. At the molecular level, gas particles are randomly colliding with each other and behaving in a turbulent fashion, yet at the larger scale the picture appears uniform and relatively simple to describe.[6]

This statistical character of physical laws reveals itself somewhat differently at different scales. And mathematical constructs that we have developed to model reality can vary in optimality, depending on the scales of interest. Thus, classical physics manifests itself quite optimally at medium (human) scales, while relativity theory emerges in scales of the very large, and quantum mechanics dominates in scales of the very small.

We will describe some aspects of these scale-associated constructs in the following sections.

3.1 Classical Physics

Near human scales (roughly in the middle row of Figure 2.1) most natural phenomena can be modeled quite well within the formalisms of classical physics. Perhaps the best known among these are Newton's laws of motion.

One aspect of classical laws is that they are expressed in decoupled coordinates of space (the familiar three-dimensional Euclidean space) and time (termed Newtonian time). Also, particles in classical systems are endowed with definite positions and velocities, both simultaneously knowable to an arbitrary degree of precision.

Classical physics essentially corresponds to the realm of our normal sensory experience, and this is why classical laws are easily visualizable and strongly intuitive. This makes perfect sense, since the human animal is biologically adapted to medium dimensions and our survival depends

---

[6][MA14, p. 52], [She15, sec. 2.1]

on developing the corresponding forms of intuition in terms of space, time, matter and causality.[7]

Nonetheless, when studying natural phenomena involving scales far away from medium dimensions, classical physics starts losing its predictive powers. Departure from classical physics becomes more apparent with the introduction of artificial sensory instruments, and the replacement of the human observer by recording apparatus.

In the world of medium dimensions, Euclidean space and Newtonian time apply as optimal and powerful abstractions. However, transitioning to different scale realms has important effects on the nature of physical laws. For example, in the realm of the very large, three-dimensional space fuses with one-dimensional time into a four-dimensional continuum, and in the realm of the very small, knowledge of position and velocity abandons its deterministic character.[8]

3.2 Theory of Relativity

The theory of relativity encompasses both special and general relativity. For the sake of simplicity and brevity, we will sketch out the basic idea behind special relativity only, as an illustration of nature's behavior in the realm of the very large.

The theory of special relativity was developed by German physicist Albert Einstein in 1905, motivated by the experimentally-verified realization that the speed of light is the same for all observers, regardless of the motion of the light source or observer. Einstein recognized that since speed is nothing but distance divided by time, space and time are not independent and absolute, but are relative and complementary. Special relativity declares that the combined speed of any object's motion through space and through time is precisely equal to the speed of light.

The theory, at its core, introduces something that is not relative, but absolute: *spacetime*. Space is not absolute. Time is not absolute. But according to special relativity, absolute spacetime (a four-dimensional continuum) forms the fabric of the universe.[9]

The relativity of space and time is a startling conclusion. Consider a parked car, for example. From your point of view the car is stationary

---

[7][Ber69, p. 241]
[8][Wie65, sec. 1]
[9][Gre07, p. 58], [Ber69, p. 226]

in space, so all its motion would be through time. But if the car speeds away, some of its motion through time would be diverted into motion through space. Because the combined speed through spacetime must be unchanged, the car's motion through time would slow down. And if the car were to reach light-speed motion through space, all the motion through time would have been diverted. A clock inside the car would stop ticking (from your point of view). However, for a different observer, say a passenger in a bus going at the same velocity as the car, things would look different. For her, the clock inside the car would tick normally, just like the watch on her wrist. The car would appear stationary, and you would be receding away. It is the same reality; it just looks different from different perspectives.

That, in a nutshell, is special relativity. The previously-assumed universality (the oneness) of time was shattered by it.[10]

Relativity places a hard limit on how fast anything can travel in the spacetime continuum, this limit being the speed of light. Nothing can travel faster than light: neither objects, nor particles, nor waves. Not even information or causality.[11]

Relativity definitely has a non-intuitive flavor. Of course this isn't surprising, since one is hard-pressed to find the survival advantage offered by a solid grasp of relativity. Our senses are under no evolutionary pressure to develop relativistic awareness. But if you think that relativity is strange, you will find that quantum mechanics is at an entirely different level (pun intended).

3.3 Quantum Mechanics

Quantum effects manifest themselves at very small scales, typically at the molecular level and below.[12] They are often described as weird and even spooky, mostly because they are so far removed from our everyday experience and the realm of our senses. But with all their weirdness, understanding of quantum laws has given us indispensable forms of technology.[13]

Without quantum mechanics' elucidation of how electrons move

---

[10][Gre07, p. 48, 128]
[11][Gre07, p. 77]
[12][She15, sec. 3.2]
[13][Dav19, p. 145]

through materials, we would not have understood the behavior of the semiconductors that are the foundation of modern electronics. And without an understanding of semiconductors we would not have developed the silicon transistor and, later, the microchip and the modern computer. It has been estimated that over one-third of the gross domestic product of the developed world depends on applications that would simply not exist without our understanding of quantum mechanics.[14]

Quantum mechanics is also at the core of other amazing technologies such as magnetic resonance imaging (MRI), which allows us to see very detailed images of soft tissue inside our bodies.

Quantum mechanics is likely to be the least familiar topic for most readers of this book. For that reason, and because we will refer to a few of these concepts later, we dig slightly deeper in the next sections.

3.3.1 Wavefunctions and Uncertainty

In January 1926 the previously mentioned Erwin Schrödinger wrote a paper outlining a picture of the atom, where subatomic particles were characterized as waves spread throughout it.[15] In this paper he introduced what is known as *Schrödinger's equation*.

Schrödinger's equation is considered a fundamental equation (not derived). It is accepted on the basis of the copious predictions it makes that have remarkable agreement with experiments.

Schrödinger's equation is a partial differential equation[16] that dictates the time evolution of what is called a *wavefunction*. A wavefunction specifies the complete state of a system of particles at any point in time, and is of the form:

$$\Psi(\mathbf{r}_1, \mathbf{r}_2, ...\mathbf{r}_n, t) \tag{3.1}$$

where $\mathbf{r}_1, \mathbf{r}_2, ...\mathbf{r}_n$ represent variables such as position vectors of fundamental particles (like electrons and protons). If $\mathbf{r}_i$ describes particle position, then the wavefunction is associated with the joint probability that particle 1 is at $\mathbf{r}_1$, particle 2 is at $\mathbf{r}_2$, and so forth, at time t.[17]

Schrödinger's equation does not provide wavefunction Ψ in explicit form. Rather, it places constraints that the wavefunction must satisfy.

---

[14][MA14, pp. 7-8], [Gre07, p. 436]
[15][MA14, p. 46]
[16]The actual expression is not critical for this discussion.
[17][She15, sec. 3.1]

Solving the equation means finding an explicit form of the wavefunction, which normally depends on particular conditions of the problem at hand.

For instance, we could be interested in the behavior of a particle traveling within a potential energy field. In this case, one of the relevant conditions is that the particle has a potential energy $U(\mathbf{r})$ which depends on the particle's position within the field. Making a few mathematical manipulations allows the equation to be put in the form of an eigenvalue problem, which is characterized by giving rise to a discrete set of solutions. In terms of the potential energy U, the result is that the particle can only adopt certain allowed values, separated by "quantum jumps".[18] In classical physics a particle's potential energy can vary continuously, but in quantum mechanics it must choose from a set of discrete allowed values.[19]

As we mentioned above, a wavefunction could be associated with the probability of finding a particle, such as an electron, at a certain location. But what does that really mean? Some physicist describe the electron as popping in-and-out of existence throughout its orbital shell around an atomic nucleus, not choosing a particular location unless compelled to do so via interaction with other particles. When this happens, the event is called *wavefunction collapse* (physicists say that a "measurement" has taken place).[20] Others prefer to think of the electron as being smeared over a region of space, having a "blurred" or "fuzzy" existence over this region.

The truth is that there is no universal consensus on what quantum mechanical probability waves truly represent.[21] It is difficult for us to picture this kind of behavior, because at human scales these probability distributions appear narrowly peaked by comparison, such that we can say "exactly" where a large-scale object is located. But when you descend to micro-scales, the distributions look much broader.[22]

In quantum mechanics, subatomic particles just like to hover in quantum limbo, in a fuzzy, amorphous, probabilistic mixture of all possibilities. It is only when measured, that one definite outcome is selected from the many.[23]

---

[18][She15, sec. 3.1]
[19][Sch44, p. 49]
[20][Gre07, p. 118]
[21][Gre07, p. 90]
[22][She15, sec. 3.1]
[23][Gre07, p. 112]

3.3.2 Entanglement and Nonlocality

There is a prediction in quantum theory called *entanglement*, in which two particles function as a system governed by a binding wavefunction. The theory predicts that it is possible to maintain this quantum connection even if the particles become separated by large (even extremely large) distances. This means that, by virtue of their past, such widely separated objects are committed to behaving in a coordinated manner.[24]

This prediction was particularly bothersome to Einstein, who in 1934 co-authored a paper (known as the Einstein-Podolsky-Rosen paper, or EPR) with two other physicists, aimed at exposing some presumed fundamental shortcomings of quantum mechanics.

To expand on the phenomenon of entanglement, let us refer to a quantum property called *electron spin*. This is not a spin in the sense of macroscopic solid objects, so let's just say that it is a property that can take one of two states: spin-up or spin-down.[25] And similar to the wavefunctions's positional "blur" discussed earlier, an electron remains in a blurred mix (called superposition) of spin-up and spin-down, for as long as the wavefunction is uncollapsed.

A pair of electrons can become entangled via an arrangement called *spin singlet state*, in which case the electrons are forced to have opposite spin at all times (even while in blurry superposition of spin states). If the electrons are separated (by any distance) and a spin measurement is performed on one of them, observation of the first electron causes the common wavefunction to collapse, and both electrons immediately take on specific spin values (the second electron without even being touched). Because of the entanglement, the second (distant) electron spontaneously takes on the opposite spin of the first electron.[26]

According to quantum mechanics, neither electron has a definite spin state before the measurement. However, Einstein did not see it that way. In his view, correlated states such as in the description above were determined long before the particles became separated. He thought they were simply revealed at the time of measurement.

Einstein argued that if measurement of one particle were to somehow cause a change in the other, there would have to be a delay before that could happen. This delay would have to be at least as long as the time

---

[24][Pen+17, ch. 1], [Gre07, pp. 114, 122, 442]
[25][MA14, p. 181]
[26][MA14, p. 186]

17

it took light to traverse the distance between the two particles, since nothing goes faster than the speed of light (according to relativity). If measurements could be made that revealed the expected particle states without the expected speed-of-light delay, all it would mean is that the states of both particles had definite values before the measurements.

So the real question came down to whether there was a way to show (experimentally) that the quantum states were truly undecided before measurement.

It took 30 years after the EPR paper first came out for a feasible experimental approach to be proposed. In 1964 Irish physicist John Bell reformulated the EPR problem into a testable statistical theorem called Bell's Inequality. However, the technology needed to undertake the required experiments did not exist at the time, so test results would have to wait for another 18 years. With the advent of proper technology, the EPR problem was finally settled in 1982 by French physicist Alain Aspect.

Aspect demonstrated by decisive experimental evidence the legitimacy of quantum entanglement. It turned out that the prediction of quantum mechanics was right.[27]

So does quantum entanglement invalidate relativity? Not quite. Einstein's skepticism was based on his assumption that something would be traveling from the first particle to the second, but that is really not the case. Nothing is traveling. Neither light, nor any form of electromagnetic waves, nor information.[28] There is no causality, in the usual sense, operating between the electrons.

Prior to quantum entanglement being proven conclusively, many physicists operated under the assumption that all laws of physics obeyed the principle of *locality*. Under this principle, an object is only directly influenced by its immediate surroundings. However, quantum mechanics, and Aspect's experiments in particular, revealed that two particles, even if separated by vast spatial distances, can behave as if they were right next to each other. As it turns out, our universe admits the property of *nonlocality*.

The true fabric of what we perceive as space and time can behave in ways that are indeed very challenging to assimilate.

---

[27][Gre07, pp. 106, 113], [Mit08, p. 133]
[28][Gre07, p. 442]

3.4   Beyond Standard Physics

The laws of physics are without a doubt the most powerful tools we have for understanding nature. And many would claim that our understanding of reality rests only upon two principal frameworks in physics: quantum theory and relativity.[29] This is largely legitimate, since classical physics (behavior at medium scales) can be seen as a mathematical approximation of either relativity (from above) or quantum theory (from below).

But this conception is not without limitations. Both theories claim to be universal, to work in all realms. However, when the mathematical constructs of relativity and quantum mechanics are used in conjunction, their combined equations break down and produce nonsensical answers.[30]

From the point of view that physical laws are abstractions that capture natural behavior at certain levels (necessarily being approximations), this mathematical incompatibility is explainable. Classical physics applies in the medium-size realm, so its mathematical formulations are close enough to the mathematical formulations of both the very large (relativity) and the formulations of the very small (quantum mechanics). But the realms at which relativity and quantum mechanics emerge are so far apart, that their current mathematical formulations are not compatible enough.

One way of resolving this incompatibility would be to refine the mathematical formulations. The current prime candidate for doing this is *superstring theory*, which in some of its versions includes realms of much smaller dimensions than those at which quantum mechanics is generally presumed to operate.[31] But proving or disproving superstring theory has not been accomplished yet, in large part because we don't currently possess the technology necessary to do it.

Another possibility is that the present formulations of relativity and quantum mechanics are actually the most suitable, and that a unified theory (at least as conventionally envisioned) does not exist. Physicists have been trying to reconcile quantum mechanics with relativity for about a century, without success. The two formulations not being relatable in a conventional mathematical way, yet being concurrently valid descriptions of nature, is something that can be understood in the context of *metamathematics*. Specifically, using Gödel's theorem.

Austro-Hungarian mathematician Kurt Gödel published a paper in

---

[29][She15, sec. 3.1]
[30][Gre07, p. 16], [She15, sec. 3.1]
[31][Gre07, p. 18]

1931 that had dramatic implications in the understanding of mathematics and logic. In it, he outlined what is known as Gödel's theorem (or Gödel's inclompleteness theorem).[32] The theorem essentially says that within consistent axiomatic systems, there exist valid statements that cannot be proved nor disproved using the rules of inference of the system. Let us elaborate on that.

Axiomatic systems work like this: you start with a set of accepted truths (called axioms, in this case some fundamental laws), and apply rules of inference (mathematical rules) to arrive at new truths (called theorems). In the context of physical laws, the implication of Gödel's theorem would be as follows: even if quantum mechanics and relativity were both formally true within the same axiomatic system (presumably, some consistent representation of nature), it may be impossible to derive one from the other (in an ordinary mathematical sense).

The problem does not necessarily lie in potential limitations or weaknesses of either theory, but could be due to an inherent limitation of formal axiomatic systems.

All scientific constructs (including the laws of physics) are models representing certain aspects of reality from a certain perspective, and as such, they are neither exhaustive nor unique.[33]

3.4.1 Diversity of Emergent Processes

Conventional physics tends to persist on methods based in calculus and probability. Much of common physics captures numerical relationships that can be represented by smooth curves, which can be seen in the current predominance of partial differential equations.[34]

But to properly understand a higher diversity of processes in nature, it is necessary to go beyond the order emerging from statistical-type mechanisms. Order can emerge from the interaction of a small number of components. Small groups of elements (in numbers much too small to involve probabilistic-type laws), can play a dominating role in the very orderly and lawful events within complex systems.[35] This is the case in the so-called "order from order" or "amplification" mechanisms observed

---

[32][Raa20], [Dav19, p. 70]
[33][Ber69, p. 94]
[34][Wol02, pp. 8, 162]
[35][Sch44, pp. 20, 80], [Ber69, p. 93]

in some systems, such as living organisms.[36]

British physicist and mathematician Stephen Wolfram has shown that certain natural behaviors are better modeled by going beyond conventional mathematical equations and using discrete rules such as those in computational programs (involving rules of any kind). Indeed, organized complexity can emerge from a surprisingly small number of simple discrete rules.[37] We will return to this shortly.

---

[36][Sch44, p. 80], [MA14, p. 57]
[37][Wol02, pp. 368, 471]

Chapter 4

Systemic Structure of Living Organisms

4.1 Thermodynamics and Information

The field of thermodynamics is a very important branch of physics, generally concerned with issues of heat and energy. Central to this theory is the concept of *entropy*, which is a measure of the disorder in a system.[1]

According to the second law of thermodynamics, entropy spontaneously increases over time within closed systems. In this regard, British astronomer Arthur Eddington wrote: "The law that entropy always increases, holds, I think, the supreme position among the laws of Nature. [...] if your theory is found to be against the second law of thermodynamics I can give you no hope; there is nothing for it but to collapse in deepest humiliation."[2]

The second law of thermodynamics characterizes a universal tendency in nature towards randomness and disorder. Order is difficult to maintain, and normally requires expenditure of energy. This is why the second law prevents the existence of perpetual motion machines. In real machines, converting energy into work involves irreversible processes. Going from order to disorder is easy (entropy wants to increase), but reversing that trend, or going from disorder to order, is an uphill battle.

---

[1][Dav19, pp. 5, 30], [She15, sec. 2.2]
[2][Dav19, p. 66]

People have thought about possible ways to circumvent this law for a while. James Clerk Maxwell, the famous Scottish scientist known for his equations of electromagnetism, envisioned a scenario that seemingly allowed for a violation of the second law of thermodynamics, known as "Maxwell's demon".[3]

Suppose you have a system of two connected chambers that contain a gas, and initially the system is in thermodynamic equilibrium. This would imply that the temperature is uniform at a macroscopic scale, but individual microscopic gas particles would typically be moving at different speeds. We can think of fast-moving particles as being "hot" and slow-moving particles as being "cold", and the temperature of the gas being characterized by the average of those motions.

Now envision a clever demon that can control a tiny door between the two chambers. As seen in Figure 4.1, the demon opens the door only if a fast (red) particle approaches from the left, or if a slow (blue) particle approaches from the right. Otherwise he keeps the door closed.

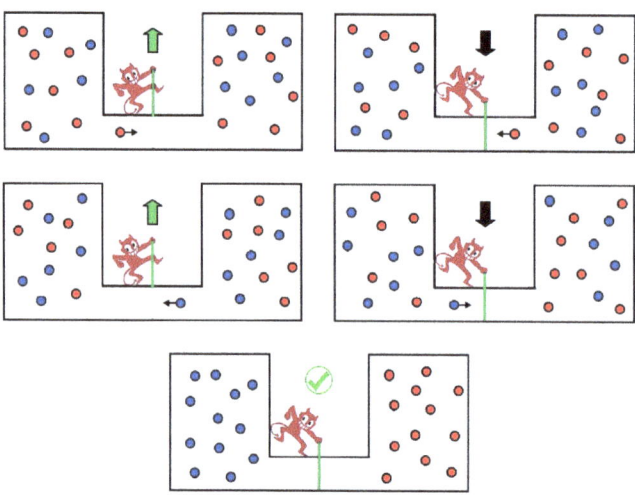

Figure 4.1: Maxwell's demon

---

[3][Dav19, p. 32], [Pen+17, ch3], [Wie65, sec. 2]

Because the particles are bouncing around randomly, Maxwell's demon is eventually able to trap most of the fast-moving particles in the right chamber and most of the slow-moving particles in the left chamber. In this thought experiment, the demon seems to be able to take the system from a state of high entropy (randomly mixed particles) to a state of low entropy (neatly separated particles into hot and cold), essentially for free.[4]

Now, it is true that the demon would need some small amount of energy to open and close the tiny door, but that amount of energy would be insignificant compared to the energy he could eventually generate by running a heat engine between the two chambers. So is the demon enabling a perpetual motion machine and violating the second law of thermodynamics?

Unfortunately for the demon and his fans, the answer is *no*. This is because in order to segregate the molecules into fast and slow categories, the demon must gather information about their locations and velocities. As it turns out, the decrease of entropy in the gas is offset by information storage in the demon.[5]

The information can be quantified, and in fact can be logically defined as the numerical negative of entropy. This concept was first proposed by Norbert Wiener of MIT (the father of cybernetics) circa 1942, and was elaborated further by another American scientist, Claude Shannon (the father of information theory).[6]

This mechanism for exchanging entropy with information is what allows for the creation of certain kinds of order in nature, without violating the second law of thermodynamics.[7] Life takes advantage of this mechanism by preserving information in the form of DNA, which as we will see later is essentially an information-storage medium.

4.2 Emergence of Complexity and Order

Living organisms are able to resist the trend towards increasing randomness via information storage, but where does the order and complexity exhibited in living things initially come from? Surprisingly enough, the

---

[4][Dav19, p. 33]
[5][Pen+17, ch3], [Dav19, p. 35], [Wie65, sec. 3]
[6][Wie65, p. 62], [Ber69, p. 152]
[7][Sch44, p. 73]

answer may lie in spontaneous self-organization processes.[8]

Temporary order can arise, for example, when water draining in a bathtub spontaneously adopts a clockwise or counterclockwise flow around the drain. Similar macroscopic order can be seen when bodies of fluid at different temperatures and/or salinity come into contact. Figure 4.2 shows such a phenomenon.[9]

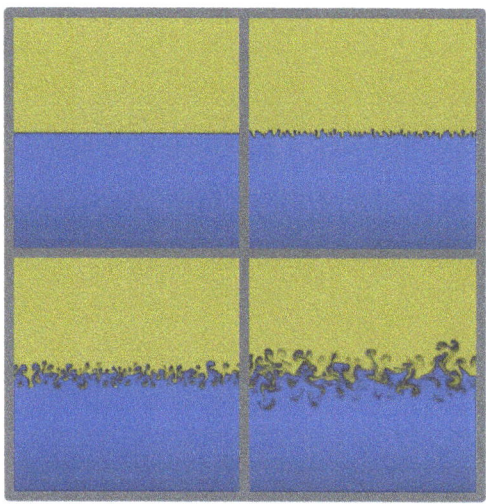

Figure 4.2: Double diffusive convection. Top layer (yellow) is warm salt water and bottom layer (blue) is cold fresh water.

The spontaneous macroscopic organization in these cases engages the participation of enormous numbers of particles (a single drop of water contains roughly 10^{21} molecules of H_2O), so this emergence of complexity might involve probabilistic effects. However, ordered complexity can also emerge from from a comparatively small number of interactions.

In this regard, Steven Wolfram performed extensive studies on what are known as cellular automata[10]. A cellular automaton is a model based on a grid of cells, where each cell has a specific state (a distinct value).

---

[8][Dav19, p. 22], [MA14, p. 299], [Wol02, p. 16]

[9]Adapted from *Double diffusive instabilities* (YouTube channel *mmnasr*, 2016); also see [Wol02, p. 377].

[10][Wol02]

The grid evolves in steps (or generations), the state of each cell in a new generation having dependence on the states of neighboring cells in the previous generation. The stepping is specifically defined by a rule (which can be of any kind).

A cellular automaton can be as simple as a row of cells with two possible states per cell (say, black or white). Figure 4.3 shows an example of such a generation-stepping rule (described below).

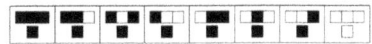

Figure 4.3: Example rule for a simple cellular automaton[11]

The top row in each frame gives one of the possible combinations of states for a cell and its immediate neighbors in the old generation. The bottom row then specifies what state the center cell should be in the new generation in each of these cases. In this example, the rule is that a cell should be black in all cases where it or either of its neighbors were black in the old generation.

Each rule of this simple type is defined by specifying one out of two possible states (black or white) for each of the eight combinations of states of three adjacent cells in the step before. That means that there are 2^8, or 256, possible rules for this kind of simple cellular automaton. If you try each one of these rules and observe the evolution of the automatons, most of the time the result is rather plain and boring. But once in a while, structure and order emerge, such as shown in Figure 4.4.

Figure 4.4 is obtained by applying the simple rule shown at the bottom left of the figure (identified as rule 110), starting with a single black cell.

Different types of rules can be considered. For example, rules that allow three states rather than two (say, black, white, and grey). The total number of possible rules of this kind is very large, but the number can be made more manageable by considering only a subset, like the so-called *totalistic rules*. The idea of a totalistic rule is to take the new color of each cell to depend only on the average color of adjacent cells in the previous generation (not specifically on individual colors). An example of such a rule is shown in Figure 4.5.

---

[11][Wol02, p. 24]
[12][Wol02, p. 32]

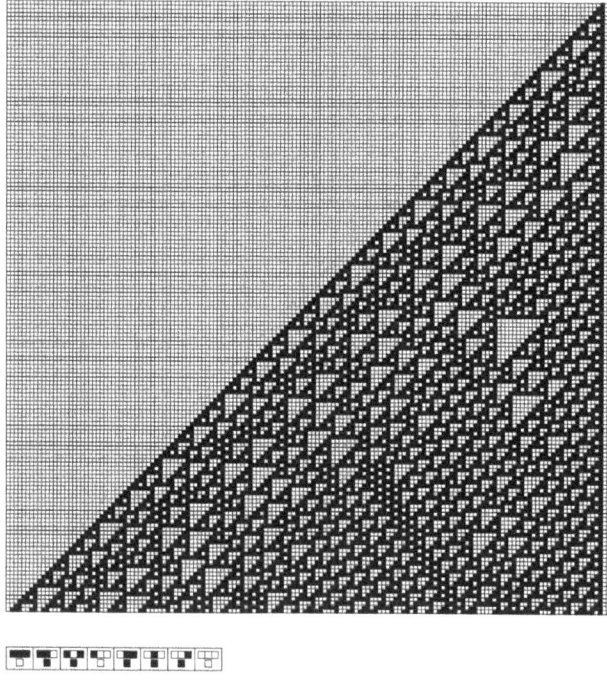

Figure 4.4: Rule 110 cellular automaton[12]

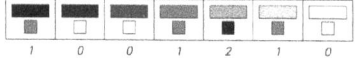

Figure 4.5: Example of a totalistic rule[13]

With three possible colors (states) for an individual cell, it turns out that there are 2187 such rules.[15] Like before, many rules produce unremarkable results, but complex order still emerges once in a while, as shown in Figure 4.6.

Wolfram called rules such as those giving rise to Figures 4.4 and 4.6 "class 4" rules. These are characterized by the emergence of structures

---

[13][Wol02, p. 60]
[14][Wol02, p. 66]
[15][Wol02, p. 60]

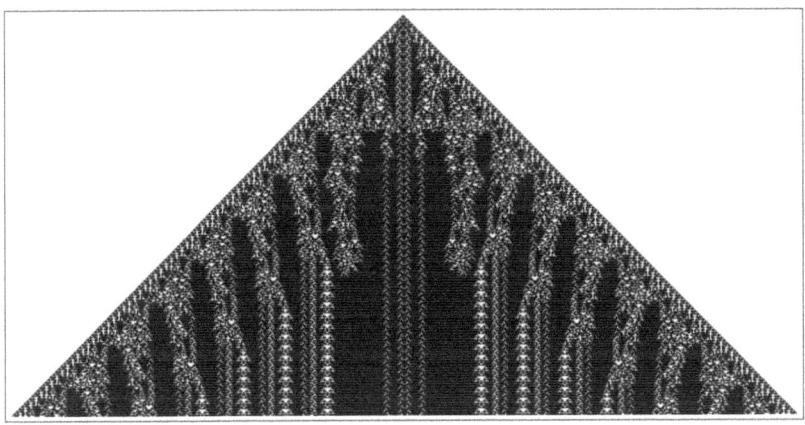

Figure 4.6: Code 1041 cellular automaton[14]

that can reappear periodically, move around, and interact with other structures in sophisticated ways.

He experimented with different kinds of automatons and different types of rules, and kept observing similar results. What Wolfram found was very unexpected and quite remarkable. His results suggest that the phenomenon of emerging complexity is universal, and independent of the type of a particular system. He also found that, while a certain threshold of complexity in the underlying rules is necessary in order for complexity to emerge in the overall system behavior, this threshold is relatively low.[16]

4.3 Persistence of Order

Ordered complexity can emerge spontaneously, but in most physical systems this order is quickly dissipated by the unrelenting effect of the second law of thermodynamics.

Schrödinger put his finger on the key issue on how life persists. For life to generate order out of disorder and circumvent the second law of thermodynamics, there had to be an entity that somehow encoded the instructions for building an organism. This entity had to be complex

---

[16][Wol02, pp. 105-106]

enough to embed a vast quantity of information and stable enough to withstand the degrading effects of entropy. Today we know this entity as DNA.[17]

DNA is a macromolecule that contains specific sequences of smaller functional building blocks called *nucleotide bases*, used by living organisms to store and propagate information. The information contained in DNA is used within cells to make sequences of aminoacids, which results in proteins. These proteins perform a variety of roles that determine the morphology and function of organisms.[18]

Because of DNA, organisms do not have to proceed by trial and error and "reinvent the wheel" at each generation. They can profit from life's past experience, in what Wiener called *phylogenetic learning*.[19] This allows for a progression of increasing levels of complexity and sophistication.

The detailed natural processes that led to the formation of DNA are not fully understood, but we are aware of some key facilitating mechanisms. In the 1920s Russian biochemist Alexander Oparin and English biologist John Haldane independently put forward what is now known as the Oparin-Haldane hypothesis. Both scientists proposed that the atmosphere of the early Earth was rich in hydrogen gas, methane and water vapor that, when exposed to lightning and solar radiation, combined to form a mixture of simple organic compounds. They argued that these compounds then accumulated in the primordial ocean to form a warm, dilute organic soup, which interacted with its environment until a random combination of its constituents yielded a new molecule with an extraordinary property: the ability to replicate itself. Oparin and Haldane proposed that the emergence of this primordial replicator was the key event that led to the origin of life as we know it.[20]

This theory is not not just based on speculation. In 1953 American chemists Stanley Miller and Harold Urey published experimental results that lent it significant support. Miller simulated the primordial environment of Earth by filling a bottle with water (to mimic the ocean) and topped it off with the gases that were thought to have been present in the atmosphere. He then simulated lightning by igniting the mixture with electric sparks. After only a week of subjecting this primordial medium

---

[17][Dav19, pp. 6, 126]
[18][Bro18, pp. 229, 312–314]
[19][Wie65, sec. 9]
[20][MA14, p. 270-271]

to sparking –to his surprise and that of the scientific community– the bottle was found to contain significant quantities of amino acids, the elemental building blocks of organic structures.[21]

Today, a majority of scientists favor the idea that before DNA was used for replication by primitive living structures, there was a different molecule that performed this process: RNA (a close cousin of DNA).

The *RNA world hypothesis*, as it came to be known, proposes that a serendipitous chemical synthesis resulted in the generation of a molecule that could both encode its own structure and make copies of itself out of the biochemicals available in the primordial soup. This copying process would initially have been very unreliable, but over the course of time those RNA replicators would have incorporated proteins that improved their replication efficiency. This would eventually lead to DNA, and to what we consider living cells.[22]

4.4 The Genetic Code

Segments of DNA capable of controlling recognizable heritable traits are called genes, and consequently DNA and related topics are studied under the science of *genetics.*

The macromolecule of DNA is formed by two strands joined in a double-helix structure, as shown in the upper portion of Figure 4.7. Each strand contains sequences of nucleotide bases, which come in four types: *A* (adenine), *G* (guanine), *T* (thymine), and *C* (cytosine).[23] The sequences of nucleotide bases are not independent across the two DNA strands, but are complementary to each other. The *complementarity rule* (or *AT/GC* rule) forces *A* to bond with *T* (via two hydrogen bonds), and *G* to bond with *C* (via three hydrogen bonds). This means that if we know the base sequence in one DNA strand, the complementarity rule allows us to predict the sequence in the other strand. For instance, for a sequence of *AAAGCT* in one strand, the other strand would have a complementary sequence of *TTTCGA.*

Within multicellular organisms such as the human body, every cell that is endowed with a nucleus contains the entire genome (the full set

---

[21][MA14, p. 273]
[22][MA14, p. 277], [Bro18, p. 415]
[23][Bro18, p. 252]
[24]Adapted from [Bro18, figs. 9.3, 9.4], [Bro18, tab. 13.1].

31

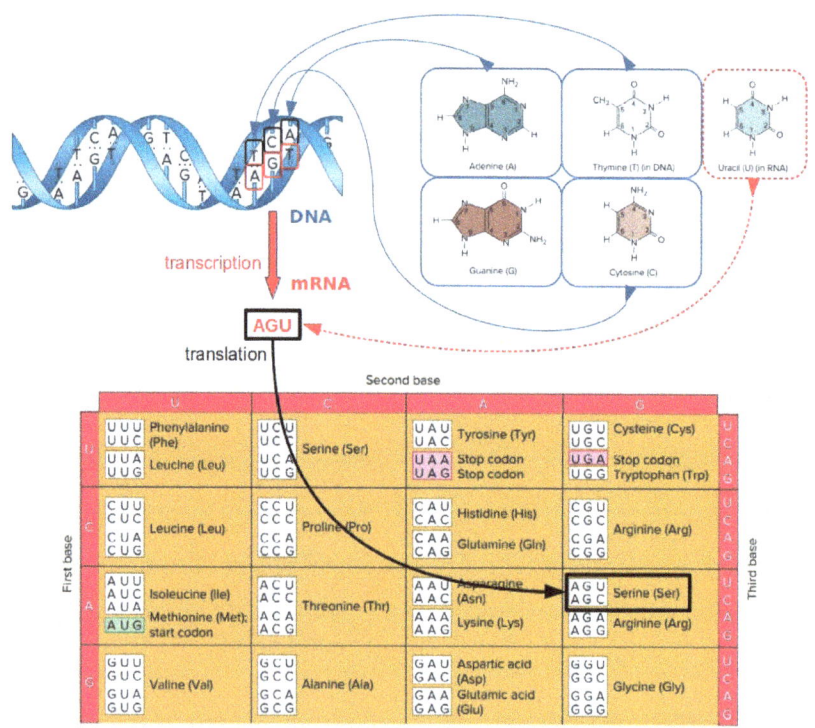

Figure 4.7: Synthesis of proteins from DNA information[24]

of DNA sequences) for that organism. This is because each cell in the organism is the result of another cell having been replicated.[25]

During the replication process, a protein called *DNA helicase* causes the two complementary strands of DNA to come apart (like opening a zipper). This exposes the nucleotide base sequences on each strand, and allows other proteins called *DNA polymerases* to synthesize two new strands (via the complementarity rule) using the existing strands as templates. Therefore, DNA is replicated in such a way that both copies retain the same information as the original.[26]

---

[25]Interestingly, mammalian red blood cells don't contain a nucleus because they eject it before reaching maturity (in order to boost oxygen-carrying capacity).

[26][Bro18, pp. 253, 259]

During cell replication, the entire DNA sequence is accessed. But most of the time, only sections of the DNA sequence get accessed for a different purpose, which is to direct the synthesis of proteins. And in practical terms, this is what DNA does. It encodes information on how to make proteins.

When accessing DNA information for the purpose of making proteins (called *transcription*), the process does not result in the creation of new DNA strands, but instead results in the generation of RNA strands.[27]

Just as in replication, the genetic sequence within DNA is used as a template, but there are significant differences. Instead of opening the DNA strands like a zipper, during transcription a protein called *RNA polymerase* synthesizes RNA within a "separation bubble", and only transcribes certain sections of DNA at a time. Also, the complementarity rule in RNA uses the nucleotide base U (uracil) instead of T (thymine).[28]

After transcription, the next step towards the creation of proteins is called *translation*. This involves the interpretation of the language of RNA (a nucleotide base sequence) into the language of proteins (an amino acid sequence).[29] The set of rules that characterizes this interpretation is known as the *genetic code* (the table on the bottom half of Figure 4.7).

Proteins are made from smaller molecules called amino acids, which are linked end-to-end forming a chain. The precise sequence of amino acids within a protein is what gives the latter its characteristic chemical properties. There are many types of amino acids in nature, but life as we know it only uses twenty of them.[30] The genetic code leverages the four types of nucleotide bases in RNA (A, G, U and C) to encode twenty amino acids, and uses unique sequences of three bases to accomplish this. These three-base sequences are called *codons*.

As an example, consider the upper portion of Figure 4.7, which shows an RNA sequence AGU resulting from the transcription of a DNA sequence TCA (because the transcribed codon is part of RNA, its third nucleotide base is uracil instead of thymine). From the table (the genetic code), codon AGU corresponds to serine, so this amino acid gets added to the chain.

Translation starts with a specific codon, called start codon. From

---

[27]Note that not all RNA sequences encode proteins. Those sequences that do encode proteins are said to be part of messenger RNA (mRNA).

[28][Bro18, pp. 280-281, 284]

[29][Bro18, p. 309]

[30][Dav19, p. 18]

there on, every three-base group defines the next codon, and the sequence terminates when a stop codon is reached. As the chain of amino acids grows it forms a polypeptide, and ultimately progresses into the three-dimensional structure of a protein.[31]

4.5   Evolutionary Mechanisms

The integrity of information contained within DNA is maintained remarkably well in living organisms. However, alterations in this information do occur, for a variety of reasons. These changes are called *mutations*. Mutations can happen spontaneously (like random errors during DNA replication), or can be caused by environmental factors (such as UV light or chemical agents).

Because existing genes tend to orchestrate a delicate balance within complex organisms, random mutations are more likely to disrupt rather than improve the organism's function. Consequently, living cells have adopted repair systems that can correct various types of DNA alterations, making mutations relatively rare events.[32]

Having said that, mutations do provide a source of genetic variation without which biological evolution would be impossible. And without biological evolution, organisms would tend to die out, because they couldn't adapt to changing environmental conditions.

Mutations can be classified as deleterious, neutral, or beneficial, depending on their effect (in the context of the organism's present environment).[33]

Beneficial mutations are far less common than deleterious or neutral mutations. But when they do occur, they enhance the survival and reproductive success of a species. Organisms that acquire advantageous genes through beneficial mutations become better adapted to their environment, making them more likely to survive and to contribute offspring to the next generation.

In contrast, deleterious mutations increase the likelihood of an organism's disease or death. And when mutations are strongly deleterious, they usually cause death early on in the life of individuals.

---

[31][Bro18, p. 312]
[32][Bro18, pp. 461, 479], [Dav19, p. 127], [Sch44, p. 41]
[33][Bro18, p. 695]

However, many mutations are relatively neutral, and do not affect the survivability of an organism very significantly. According to the *neutral theory of evolution*, populations in nature absorb mutations like a sponge and retain them in a sort of neutral reserve, thereby providing a source of variability for future environmental changes[34] (see Figure 4.8).

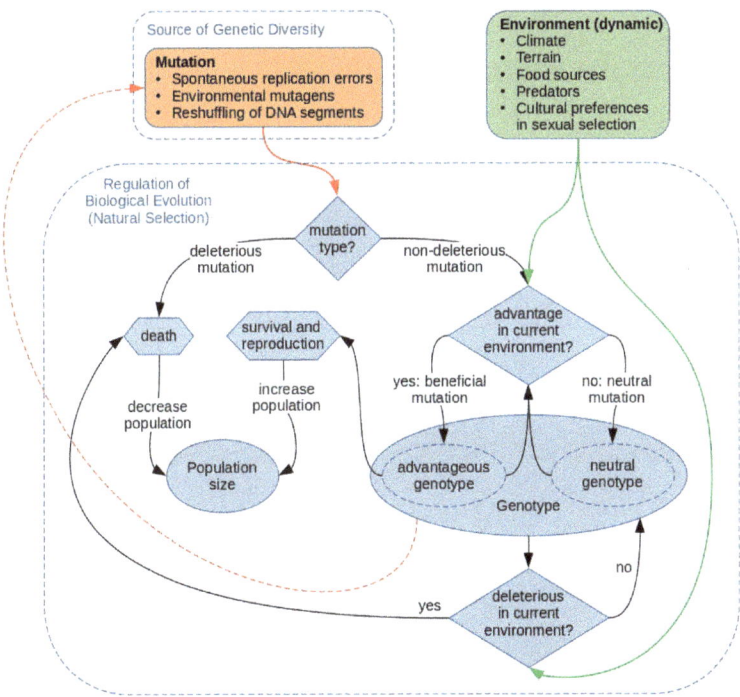

Figure 4.8: Systemic relations in biological evolution

Neutral evolution is related to the concept of *preadaptation*, in which an organism possesses certain traits that can be repurposed or adapted relatively quickly in the face of novel environmental conditions. In other words, a preadapted part is basically capable of doing a job before the job arrives. However, the concept of preadaptation does not imply that a trait arises in anticipation of filling a biological role sometime in the future.

---

[34][Bro18, pp. 695, 756]

There is no predetermined plan in evolution, just as future environmental conditions cannot be anticipated.

In a strict sense, preadapted traits must serve some immediate survival role.[35] But in a general sense, preadaptation can be thought of as a property of the neutral genomic reserve mentioned above. This reserve is formed by a progressive accumulation of non-deleterious traits, with potential of becoming advantageous.

Of course, a changing environment could just as well mean that once-advantageous or neutral genes become disadvantageous. A circumstance like this is what caused the extinction of most dinosaur species at the end of the Cretaceous period. Drastic environmental changes set off by the impact of a massive asteroid on Earth made average temperatures drop and led to food supplies being disrupted. The larger sizes and specific thermoregulatory characteristics of dinosaurs became disadvantageous, and their numbers plummeted. This allowed smaller hot-blooded mammalians to gain advantage, and eventually claim a considerable degree of dominance in the food chain.

The net result of evolution is a population that is continuously adapting to its environment. As shown in Figure 4.8, evolution is a self-correcting process in a system that tends towards dynamic equilibrium.

4.6 Gene Regulation

Animal development demands a great deal of gene regulation, which refers to the set of control mechanisms whereby genes are *expressed* (or turned on, basically to make proteins). Some genes are expressed only during early stages of development, and others only in the adult. Gene regulation is indispensable to ensure the differences in structure and function among distinct cell types (cell differentiation). For instance, nerve cells and muscle cells contain the same set of DNA instructions, yet are strikingly different in form and function because of gene regulation.[36]

One way in which genes can be influenced is via environmental factors, such as in *epigenetic effects* (environmental influences can include temperature, and even diet). Epigenetics is basically a layer of instructions that helps control how DNA is interpreted.[37]

---

[35][Kar15, p. 20 p715]
[36][Bro18, pp. 361, 645], [Kar15, p. 204]
[37][Dav19, p. 74], [Bro18, pp. 388, 391]

And beyond epigenetics, genes leverage a variety of different phenomena to help them build an organism. This is a necessity, because the information contained in DNA is not enough by itself to build a complex organism. For example, genes can rely on processes based on mechanical reaction forces and geometric rules.[38]

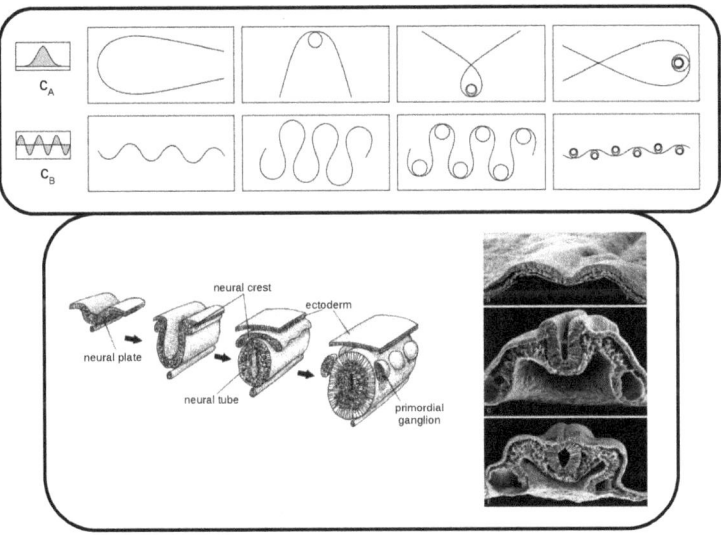

Figure 4.9: Folding processes and neurulation[39]

As an individual organism develops, different sections of its genetic program can cause multiple kinds of cellular growth to occur. For instance, in some cases uniform growth can take place, and in other cases differential growth may cause various kinds of local folding.[40] Some of the diversity and complexity that we observe in biological structures can result from surprisingly elementary effects.

The top frame of Figure 4.9 shows what happens when local curvature is made to vary according to several simple rules. Going from left to right, the maximum amplitude of local curvature (of the forms c_i as shown)

---

[38][BG12]
[39]Adapted from [Wol02, p. 418], [Kar15, fig. 16.8], [Sch+15, fig. 4-5]
[40][Wol02, pp. 417-422]

takes on increasing values along contours.[41]

The bottom frame of the figure shows drawings and computer-enhanced photographs of neurulation in vertebrate embryos, a process which includes the transformation of the neural plate into the neural tube.

Even though genes are ultimately responsible for stimulating growth in embryonic tissue via the production of chemicals that can vary in concentration over certain distances,[42] it is the geometry of mechanical tissue interactions, as opposed to genes, that are most immediately correlated to the basic morphology that results from neurulation.[43]

4.6.1   Gene Organizational Networks and Hierarchies

Body folding in the embryo is a critical event in the development process of most animals, as it establishes the tube-within-a-tube body plan that defines the animal's body axes.[44]

In general, there are four major steps in the pattern-development program of most animal species:[45]

1. formation of body axes
2. segmentation of the body
3. determination of structures within the segments
4. cell differentiation

Following the steps above, after the embryo becomes organized along axes and then into a segmented body pattern, genetic processes turn towards determining the fate of cells within segments (geneticists use the term *cell fate* to describe the morphological features that a cell or group of cells will ultimately adopt).

The genes that control cell fate in a particular region of the body are known as *homeotic* genes.[46]

Homeotic genes are master gene switches that bring under their command legions of secondary genes, which are in turn responsible for the formation of body parts. It is interesting that the shaping of bodies is controlled by a rather small percentage of genes, and that these regulatory

---

[41][Wol02, p. 418]
[42][Bro18, p. 644]
[43][BG12]
[44][Sch+15, p. 85]
[45][Bro18, p. 646]
[46][Kar15, p. 204]

genes are ancient, shared by all animals. The top of Figure 4.10 depicts the organization of homeotic genes in Drosophila (fruit fly), laid out in two clusters along one of its chromosomes.[47]

Note that the anterior-to-posterior order of homeotic genes along the chromosome correlates with the order of expression in the embryo, and that the same organization can be observed in the adult as well.[48]

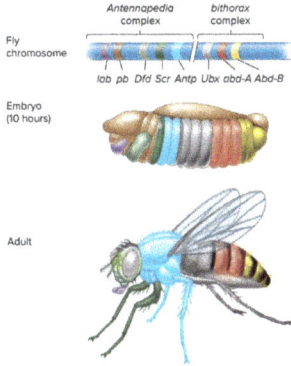

Figure 4.10: Expression pattern of homeotic genes in Drosophila[49]

There are groups of homeotic genes in vertebrate species that are homologous to those in insects, called Hox complexes.[50] Although mammalian Hox genes have been individually altered through evolution, they retain significant sequence homology to insect homeotic genes. The amino acid sequences encoded in mammalian Hox genes and those in corresponding Drosophila homeotic genes are about 90% identical.[51]

Hox genes are capable of regulating many structural genes (often a hundred or more), which in turn make products involved in building specific body structures. Consequently, even one small change in a Hox gene can magnify into huge effects throughout the downstream structural genes over which it presides. A change in a Hox gene can add segments,

---

[47]A chromosome is a DNA macromolecule.
[48][Bro18, p. 654]
[49]From [Bro18, fig. 26.12].
[50][Kar15, p. 204], [Bro18, p. 659]
[51][Sch+15, p. 124]

or legs, or wings, or remove them.[52] Hox genes are essentially high level programs within a hierarchy of programs that encodes rules on how to build living organisms.

Figure 4.11 illustrates Hox gene expression in the human arm, over time and location.

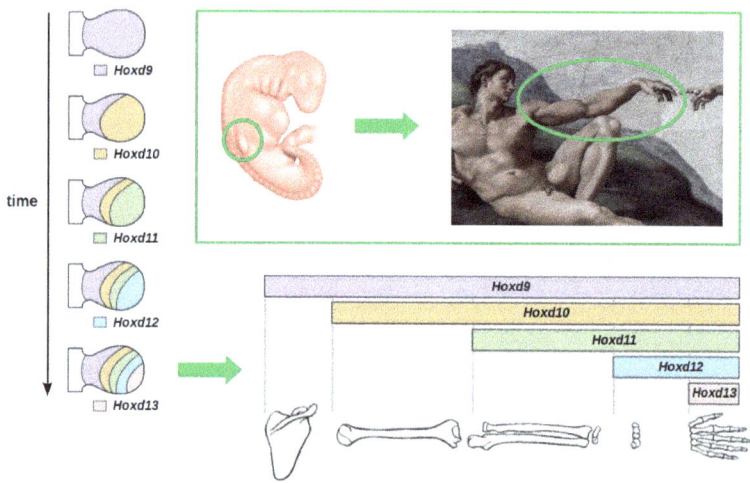

Figure 4.11: Hox gene expression in the human upper limb[53]

4.7   Structural Adaptation of the Body Plan

Evolution proceeds most often by remodeling of a basic underlying plan, and not by all new construction. Commonly, a new part is just an old part modified for new purposes. As mentioned earlier, if a complete novelty made a sudden appearance, it would probably disrupt the organism's functional harmony and would likely be selected against.[54]

---

[52][Kar15, p. 204], [Bro18, p. 661]

[53]Adapted from [Sch+15, figs. 20-8, 20–9], *Fetal development four weeks after conception* (www.mayoclinic.org) and *Creazione di Adamo* by Michelangelo.

[54][Kar15, pp. 21, 203]

The present section highlights the fact that our current morphology depends to a large extent on previous states that we have experienced as part of our continuous adaptation to changing conditions. As a case in point, we will focus on locomotive adaptations.

4.7.1   Evolution of the Human Skeletal Structure

The most primitive form of locomotion in vertebrates is that of fish, where sideways undulation is used to generate forward propulsion.

As vertebrates radiated out of water and onto land, their morphology underwent significant changes. The reasons for why some vertebrates moved to dry land are not completely clear, though there are a few plausible scenarios. Fish that venture onto land today, such as the mudskipper, apparently do so in search of food and to leave behind water-bound predators. Similar advantages may have favored the move of vertebrates with the right preadaptations (such as the "walking fins" of modern lungfish) onto land, thereby giving rise to the terrestrial phase of vertebrate evolution.[55]

Figure 4.12: Transition of locomotion from water to land[56]

The wavy lateral undulations inherited from swimming were carried forward from ancient fish into early *tetrapods*, forming the base of a modified mode of locomotion adapted to walking on land.[57]   Just like

---

[55][Kar15, pp. 347, 447]
[56]Adapted from [Kar15, figs. 8.24, 9.28].
[57]Tetrapods are animals with four limbs (including humans).

modern reptiles, early tetrapods used lifting and planting movements of the limbs to establish points of support and traction, which were synchronized with the lateral undulations of the spinal column (see Figure 4.12).

As locomotion became optimized for more rapid and sustained transport on land, limbs changed from a sprawled position to an under-the-trunk position, and flexing of the vertebral column transitioned from primarily side-to-side (on a coronal plane) to predominantly up-and-down (on a sagittal plane).[58] Accompanying these adaptations, increased dynamic efficiency was brought about by morphological (structural) twisting of the humerus and femur, pointing the digits forward and thus placing them in line with the direction of travel, as shown in Figure 4.13.[59]

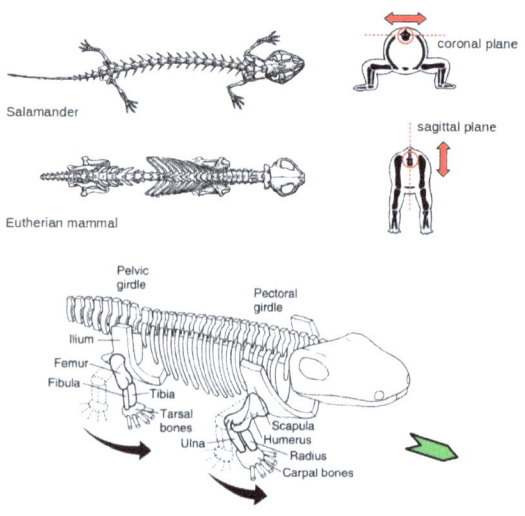

Figure 4.13: Terrestrial locomotion adaptations[60]

Along with adaptations that relate to dynamic aspects of locomotion, there are additional structural traits in tetrapods that have a clear static-loading origin.

---

[58]A coronal plane divides the body into a dorsal part and a ventral part, while a sagittal plane divides the body into a left part and a right part (see Figure 4.13).

[59][Kar15, pp. 352-353]

[60]Adapted from [Kar15, figs. 9.32, 9.33].

Before making their transition to land, vertebrates exhibited a rather straight spinal column (mostly flat on a horizontal coronal plane). The reason is that in an aqueous medium, buoyancy forces are evenly spread along the body. Contrasting this, in quadrupedal tetrapods only two pairs of points (the forelimbs and the hindlimbs) provide support. As we illustrate below, the long span between support points that characterizes quadrupedal stance is responsible for the dorsally-convex thoracic curvature that can be observed in the spinal columns of many modern tetrapods (not just quadrupeds).

As shown in Figure 4.14, there is substantial homology in the static gravitational loading of a quadruped and that of a tied-arch bridge. In a tied-arch bridge, the arch structure is predominantly loaded under compressive stresses, while the tie beams under the deck experience mostly tensile stresses.[61] Similarly, in the "arched" quadruped the weight of the torso causes compressive stresses on its dorsal aspect (vertebral column) and tensile stresses on the ventral aspect (abdominal muscles and the sternum).

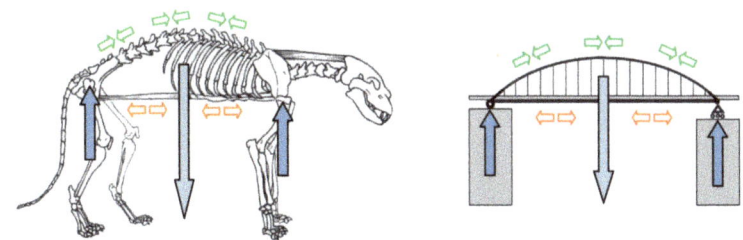

Figure 4.14: Static loading in quadruped and in tied-arch bridge[62]

The dorsally-convex thoracic curvature optimizes the structural geometry by reducing peak stresses, and loads the spinal column preferentially in compression. Notably, this latter effect leverages fundamental anatomical features of the spinal column, which evolved under repeated compression forces (imposed by locomotive musculature) inherent in primordial undulatory movements.

Structural stresses resulting from basic geometry is something that

---

[61][BLP16]
[62]Adapted from [Kar15, fig. 8.35] and [BLP16]

both natural evolutionary processes and human ingenuity have had to deal with. So it is not surprising that nature and good engineering have arrived at similar solutions for similar problems.

Humans inherited this dorsally-convex spinal curvature from quadrupedal tetrapods, and later acquired an additional dorsally-concave curvature just below it (lumbar curvature), as they stood up and became bipedal.[63]

Incidentally, humans acquired other notable structural features originating from a very different kind of locomotive environment (think of trees as opposed to ground). Human ancestors include tetrapods who evolved a form of locomotion called *brachiation*, where individuals swing between tree branches using their arms (like gibbons today). This gave humans relatively long forelimbs and grasping hands. Among our inherited brachiation adaptations is also a very prominent clavicle,[64] which increases our freedom of motion at the base of the arm.

The human machine's structural design bears the clear signature of evolutionary processes in our past. The convex and concave curvatures of our spine, the structural twist in our arms and legs, our feet pointing forward, our long arms and our grasping hands. All are due to an evident sequence of adaptational events (see Figure 4.15).

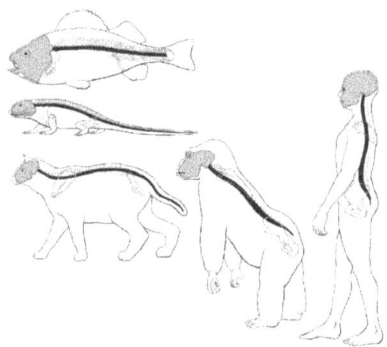

Figure 4.15: Evolutionary adaptation of the skeletal structure[65]

---

[63][Dim11, ch. 4]
[64][Kar15, p. 354]
[65]From [Dim11, fig. 4-5].

4.8 Amplification Mechanisms

The European robin (Erithacus rubecula) is a small songbird that possesses a very interesting means of navigation. It has the ability to sense the direction and strength of the Earth's magnetic field, in a fashion akin to an inclination compass.

An inclination compass basically works by measuring the angle between the local surface of the Earth and the direction of the Earth's magnetic field lines. It can distinguish between travel towards the equator or towards a pole (based on whether the angle is increasing or decreasing), but it cannot distinguish between north and south poles specifically (since the field lines make the same angle with the ground regardless of the hemisphere).

But this is not the interesting part. What is intriguing about the robin's ability is that it can detect the Earth's magnetic field in spite of the latter being exceedingly weak. There are no known biochemical processes that would allow an animal to detect the earth's magnetic field, given the levels of energy available to effect chemical reactions. At least not within the realms of conventional biochemistry.[66]

Amazingly, there is a good amount of evidence that the European robin leverages the quantum effect of entanglement (which we discussed in section 3.3) to make its navigational system work. The key to the robin's magnetoreception abilities lies in a process known as the *entangled radical pair mechanism*.[67]

This mechanism relies on the presence of a light-receptor protein called cryptochrome, which is found in the eye of the robin. When cryptochrome is exposed to daylight, it can form pairs of free radicals (molecules that have lone electrons in their outer electron shell) that are quantum entangled. A pair of entangled electrons forms a superposition of quantum spin states that can be *singlet* (electrons have opposite spin) or *triplet* (electrons have the same spin state). The balance between the singlet/triplet states is exquisitely sensitive to the direction and strength of external magnetic fields, so that the direction in which the bird flies influences the dynamics of chemical reactions that affect the activation of cryptochrome. Activation of cryptochrome in turn affects the light-sensitivity of retinal neurons, with the overall result that the

---

[66][MA14, pp. 5-6]
[67][MA14, pp. 181, 190–194]

magnetic field is perceivable by the animal.[68]

The idea that quantum effects can influence biological systems in any significant way is quite surprising. This is because quantum effects such as entanglement only become apparent under *coherence*, which is very delicate and extraordinarily difficult to maintain.[69] In fact, decoherence is one of the fastest and most efficient processes in the whole of physics. Just think about all the material surrounding any given particle, constantly interacting with that particle and effectively "measuring" it, thus collapsing its wavefunction and forcing it to behave as expected in classical physics. When decoherence happens, the exotic nature of quantum probabilities snaps into the more familiar processes of classical phenomena. So the notion that coherence can be maintained in the hot, wet and molecularly turbulent environment inside a living organism is rather extraordinary. And yet, the sophisticated mechanism within the robin's magnetoreceptor system is able to pull it off.[70]

The intent of this discussion about magnetoreception is not intended to focus on quantum mechanics in particular. It rather serves to highlight the fact that nature and evolutionary processes can take advantage of all sorts of unusual and unexpected tricks, in order to help organisms adapt to their environments and needs.

However, the mention of quantum mechanics is relevant in that the robin is able to leverage effects that normally remain in the realm of the ultra-small and ultra-fast, and makes them usable at a completely different scale.

An interesting lesson of magnetoreception is that fairly small events can have large consequences.[71] Magnetoreception is an extreme example, but a useful one in illustrating what are known as *amplification mechanisms*.[72]

An amplification mechanism is one in which a small change in a "leading part" of the mechanism can cause considerable (even dominating) changes in the behavior of the total system. Amplification is common in systems of all kinds, and is one way in which a hierarchic order of parts or processes may be established.[73]

---

[68][MA14, pp. 14, 181, 191]
[69]Coherence means that the wavefunction has not collapsed.
[70][Gre07, p. 210], [MA14, pp. 117, 126]
[71][MA14, p. 188]
[72]The workings of Hox genes discussed in section 4.6 exhibit a similar kind of effect.
[73][Wie65, secs. 3, 7, 10], [Ber69, pp. 71, 95, 117, 214]

4.9 Neurons and Networks

Perhaps one of the most basic instantiations of an amplification mechanism is the *relay*, which is nothing more than a switch where a low-power
electrical signal is used to open or close the flow of a second electrical
signal. It can be used as an amplification device (if the second signal is
of larger power characteristics than the first), but it is also commonly
used as a networked switching element (within networks of switching
elements).

A sophisticated version of the relay that is used in modern electronics
applications is the *transistor*. In this device, semiconductor materials
allow an electric current applied at one pair of the device's terminals to
alter the conductivity at a second pair of terminals, and thus control
the current flow through it. The transistor is what enables modern
high-density integrated circuits, such as microprocessors and memory
chips. It is, in fact, the basic building block of our computer-based era.

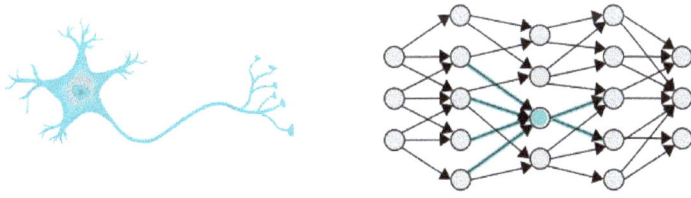

Figure 4.16: Neuron and neural network[74]

It is notable that animal nervous systems, which are at a high level
akin to artificial computation systems, contain elements that behave like
transistors.[75] We are, of course, talking about neurons, or nerve cells.

But to be fair, a neuron is probably more analogous to an aggre-
gation of multiple transistors. Neurons have an obvious morphological
feature that reflects their multi-port communication capability: extensive
branching.[76]

---

[74]Neuron image by David Baillot of UC San Diego (www.jacobsschool.ucsd.edu).
[75][Wie65, sec. 5]
[76][Pur+08, p. 5]

47

The basic structure of a neuron is shown on the left of Figure 4.16, and an illustration of how it is networked with other neurons is portrayed on the right.

The right of the figure is actually a diagram of an artificial neural network (ANN), and it is no coincidence that it has substantial homology with the way biological neural circuits work. Artificial neural networks were originally inspired by their biological counterparts, but as research in ANNs and the related field of artificial intelligence (AI) evolved, new knowledge cross-pollinated with neuroscience and actually helped to advance the latter.[77]

A characteristic switch-like property of the biological neuron is that it will not propagate a signal to downstream neurons until the sum of all incoming signals exceeds a given threshold. Only after the threshold is exceeded does an electrical signal called *action potential* travel down its body and propagate to other neurons.[78] Artificial neural networks incorporate a homologous behavior in the form of an *activation function*, such as that shown in Figure 4.17.

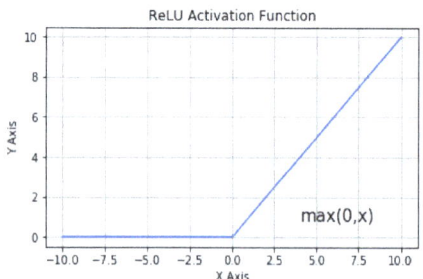

Figure 4.17: Rectified linear unit (ReLU) activation function[79]

The activation function shown in Figure 4.17 is called ReLU (for rectified linear unit).[80] Prior to widespread adoption of this activation function, other functional forms such as the logistic sigmoid function (arguably more complex, but based on probability theory), were preferred.

---

[77][MWK16, sec. 1], [Sto19, p. 135]
[78][Pur+08, pp. 7, 50]
[79]Note that in biological neurons the intensity of a stimulus is encoded in the frequency (not amplitude) of action potentials [Pur+08, p. 27].
[80][Sto19, p. 144]

But remarkably, this simpler abstraction led to substantial improvement in the capability of deep artificial neural networks (ANNs with large numbers of layers, shown as columns of neurons in Figure 4.16).

It is important to note that the systemic behavior of biological neural networks does not simply depend on connection topology and action potentials. Neurons are of an electrochemical nature, and diffusive chemical effects come into play at the interfaces between neurons.

Up until the late nineteenth century most physiologists believed that nerve cells were fused directly (at the protoplasmic level) to their neighbors, forming a continuous nerve cell network or *reticulum* (Latin for "net"). This was until Spanish anatomist Santiago Ramón y Cajal (the father of modern neuroscience) replaced this "reticular theory" with the *neuron doctrine*. He concluded, based on detailed microscopic examination of nervous tissue, that nerve cells are discrete entities and that they communicate with one another by means of specialized contacts, later called *synapses*. Although today we know that there are some synapses that permit direct passive flow of electrical current from one neuron to another, the most abundant type of synapse in the nervous system is (by far) chemical.[81]

A key feature of chemical synapses is the presence of small organelles called synaptic vesicles within the presynaptic terminal. These spherical organelles are filled with one or more *neurotransmitters*, which are chemical agents secreted from the presynaptic neuron. Neurotransmitters evoke downstream electrical responses by activating proteins in the postsynaptic neurons, called neurotransmitter receptors.[82]

Multiple neurotransmitters can produce different types of responses on individual postsynaptic cells. For example, a neuron can be excited by one type of neurotransmitter and inhibited by another. The speed of postsynaptic responses produced by different transmitters also differs, allowing control of electrical signaling over different time scales. More than 100 different agents are known to serve as neurotransmitters, allowing for tremendous diversity in the chemical signaling between neurons.

Complementary Chemical Signaling

But chemical signaling is not limited to the vicinity of synapses. Other well-characterized means of chemical communication include endocrine

---

[81][Pur+08, pp. 3-4, 7, 85]
[82][Pur+08, pp. 88, 119]

signaling, which acts via secretion of hormones into the bloodstream and is able to affect targets throughout the body.[83]

Biochemical systems have a number of characteristics that make them complex and sometimes difficult to understand. They typically have a large number of distinct chemical species, which can combine and interact in intricate and sophisticated ways. In addition, the dynamic nature of biochemical systems (changing behavior over time) is something that cannot be ignored.[84]

4.9.1 Hierarchical Network Structures

The human nervous system is formed by vast networks of neurons, exhibiting both physical and functional structure. Physically, the nervous system encompasses the brain, the spinal cord, and nerves.

Given that the human brain is estimated to contain roughly 100 billion neurons,[85] the need for efficiency in structuring should be evident. As it turns out, the brain is organized into modules capable of performing discrete computational tasks.[86]

A fundamental objective of the nervous system is to ensure survival, and it could be argued that primordial components of the nervous system are built around life-saving reflexes. For example, a fly is difficult to swat because there is a quick-response loop between its wide-angle vision and its motor response for flying. Similar circuitry allows the crawfish to escape from its predators, minimizing the time between the appearance of a threatening stimulus and the potentially life-saving motor response.[87]

Mammals have more complex nervous systems than insects and crustaceans, but recall from section 4.7 that evolution tends to modify or build on top of existing structures, rather than create completely new ones. Various reflex-reaction circuits are evident in our bodies, like that exercised in the *knee-jerk reflex* test that physicians perform to verify the basic integrity of a patient's nervous system.

The basic modularity that supports reflexes carries over into more sophisticated behavior. When you perform complex actions, such as moving inside a crowded room, you make use of multiple sets of control

---

[83][Pur+08, p153]
[84][II10, p. 265]
[85][Pur+08, p. 10]
[86][MWK16]
[87][Pur+08, p. 87]

modules, built one on top of the other in a hierarchical structure. At a high level you may set the goal to grab a cup of coffee at the opposite end of the room. From there, a lower level computational module deals with planning a navigation path that avoids furniture and standing people. And then an even lower level module figures out how to move your legs to follow that path, leveraging local feedback control by means of the spinal cord to execute the motions.

Looking at the brain, we find a number of discrete specialized structures, such as the cortex, the thalamus, hippocampus, basal ganglia and cerebellum. These structures are different both anatomically and functionally, and have been found to deal with distinct computational problems. The cerebellum, for example, uses deviations between our planned movements and the movements that we actually execute, to adapt to properties of the environment. On the other hand, areas like the thalamus and hypothalamus appear to have information from other areas flowing through them, arguably performing information-routing tasks. Even within the much younger cortex (phylogenetically speaking), subregions are highly specialized into distinct areas that deal with sensing, planning, motor actuation, processing of language, emotion, etc.[88]

4.10 Emergent Abstractions

From a systems perspective, the animal nervous system can be seen as a set of hierarchical networked modules (underpinned by neural circuitry), subject to modification by "chemical environment" mechanisms. This systems-mechanistic network arrangement is what gives context to the self-regulating control loops that are evident in the autonomic system (which we will discuss in Chapter 7).

Many details regarding biological organisms are currently known, but traditional models have not yet comprehensively integrated all the kinds of complexity we see in biology, especially in terms of general theories.

In the presence of that complexity, using the right abstractions might allow for the emergence of relatively simple relationships and behaviors, capable of integrating key aspects of this knowledge. These could capture some of the essential mechanisms that we observe in biological forms and processes. Under the right lens it becomes possible to make a wider range of models for biological systems, and potentially to see how to emulate

---

[88][MWK16, secs. 1.3, 4]

the essence of their operation. And using general principles in simple models, it might be possible to construct new kinds of general abstract theories in biology.[89]

---

[89][Wol02, p. 8]

Part II

Application of Systems Thinking

Chapter 5

Case Study

Humans are curious by nature, and have been tinkering with tools and mechanisms (natural and man-made) for at least as long as we've been *homo sapiens*. We are very fast learners, and have a well-developed capability to pass on cultural knowledge from generation to generation.

What we have learned by tinkering, especially if useful or advantageous, is something we have tried to preserve. And among the useful things that we have learned, there are some that apply to recovering from pain or disease (for "maintenance of the human machine", as it were). A particularly interesting body of knowledge that falls into this category is what can be described as *contact therapies*.

Contact therapies are methods that rely on physical stimulation of distinct locations on the body surface, aimed at relieving certain conditions. Among these, perhaps the best known is acupuncture.[1]

5.1 Acupuncture

Acupuncture is a form of therapy in which needles are used to stimulate specific points on the body surface, in order to relieve targeted conditions. In spite of its use of needles, acupuncture is fundamentally non-invasive.

---

[1] An additional example is *Tok Sen* (a traditional technique practiced in Thailand, where a blunted stake is repeatedly tapped with a mallet to stimulate particular points on the body). Another is *shock wave therapy* (a novel method that is becoming increasingly popular in orthopedics and traumatology).

This is because treatment is mostly restricted to the skin, and the needles themselves are quite thin. Additionally, acupuncture techniques can be applied without any needle penetration at all. Stimulation of therapeutic points –or *acupoints*– can be done in the form of acupressure (applying pressure with fingertips or other blunt objects), electric acupuncture, and even laser acupuncture (photo-acupuncture).[2]

The oldest written records of acupuncture as an organized system of diagnosis and treatment can be traced to ancient China, going back some 2500 years.[3] But there are indications that knowledge of acupoints could be much older than that. The frozen remains of a human male (colloquially called Ötzi) that lived near the Italian Alps about 5300 years ago, was found to a number of unusual non-decorative tattoos on his body. Based on tattoo locations and the musculoskeletal conditions of this individual (deduced from spinal imaging), there is evidence that the tattoo patterns reflect an acupuncture-like treatment for pain.[4]

Acupuncture is sometimes thought of as whimsical Eastern medicine, lacking any scientific footing. However, this is an unfounded misconception. Both the National Institutes of Health (NIH) and the World Health Organization (WHO) have reported scientific consensus that there is positive, incontrovertible evidence for the effectiveness of acupuncture in a number of conditions.[5]

The problem for acupuncture, within the context of today's scientific *Zeitgeist*,[6] is that, even though compelling effects can be observed, there is no clear-cut mechanism able to convincingly explain them.

In its current state, acupuncture is probably best described as a *black box* mechanism. Inputs and outputs are reasonably-well established, but exactly how things work is rather murky.

Given some unique advantages of acupuncture, there is good reason to pursue a better understanding of this kind of treatment. For example, reports of controlled clinical trials show that acupuncture is effective in the treatment of rheumatoid arthritis, although with a lesser strength than corticosteroids.[7] But because acupuncture treatment does not cause serious side-effects (corticosteroids do), it is advantageous in many cases

---

[2][Org03]
[3][Hea97]
[4][KT+13]
[5][Org03], [Hea97]
[6]German expression meaning "spirit of the times".
[7][Org03]

to use acupuncture for treating this condition. Another example is the treatment of kidney stones. Although the success rate of acupuncture therapy in treating kidney stones is by no means as high as that of surgical intervention, its simplicity and reduced risk might make it a good choice.[8]

The way in which people acquired knowledge about acupoints is speculative, but it was likely the result of observation, and trial and error. Presumably, people noticed the existence of tender spots on the body during the course of certain diseases, which when stimulated by contact or heat resulted in the alleviation of symptoms. This knowledge was gradually accumulated, and consistency in certain patterns led to the discovery of acupoints. The credibility of this scenario is supported by the work of English neurologist Henry Head, who in the late 1800s discovered patterns of points closely resembling traditional acupoints, by a trial-and-error process similar to the one just described (we will revisit this in the next section).

In general, acupoints appear to be located near richly innervated and vascularized structures such as:[9]

- cutaneous nerves emerging from deep fascia
- large peripheral nerves
- motor points of neuromuscular attachments
- bifurcation points of the peripheral nerves
- ligaments rich in nerve endings
- blood vessels in the vicinity of neuromuscular attachments

These appear to roughly characterize the "sensory" aspect of the acupuncture mechanism.

In terms of the "processing" aspect, things are much less clear. However, there are strong indications that acupuncture leverages the topology of spinal nerves as well as the circuitry and reflexes of the autonomic nervous system, as we will see later.[10] Having said that, definite explanations of the mechanism (or mechanisms) of acupuncture remain elusive. But if we put acupuncture under the lens of systems theory, traditional mysterious explanations become less necessary, and an intuition

---

[8][Org03]
[9][ZB14]
[10][Wan10, sec. 1.2]

for plausible mechanisms consistent with modern scientific thought is obtainable.

As mentioned earlier, one of the sweet spots for systems theory is the exploration of "no man's lands" of science, which have shown to be resistant to conventional methods of study. Because of this, in the context of its therapeutic effectiveness, acupuncture is a very compelling (and intriguing) case study for systems theory.

Chapter 6

Segmental Anatomy and the Autonomic System

6.1 Segmental Anatomy

Even though segmentation in vertebrates is less obvious than in invertebrates, it is definitely present.[1] For instance, the vertebral column is a segmental structure. Similarly, the lateral body musculature (such as intercostal muscles between a given pair of ribs) is structured into segmental blocks, and nerves and blood vessels supplying them also follow this segmental arrangement.[2] Among the different expressions of segmentation in the human body, a particularly interesting one is the sensory innervation of skin.

Segmental structures in the human body are generally organized following a pattern established by a series of transient embryonic condensations known as somites.[3] Somites start forming after the cranio-caudal axis of the developing embryo is established from neurulation (mentioned in section 4.6). The creation of these structures involves rythmic oscillations of a network of genes between permissive and inhibitory states, in a

---

[1]In the current context, segmentation is the division of certain body structures into a series of repetitive segments.
[2][Kar15, pp. 17-19]
[3][Sch+15, p. 175], [Kar15, p. 301]

coordinatory framework called the *clock and waveform model*.[4] Somites resulting from this process can be seen in the left frame of Figure 6.1.

Nerve fibers start to emerge from the spinal cord of the embryo around day thirty, initially leaving the spinal cord as a continuous broad band, but then condensing to form discrete segmental spinal nerves, as shown on the right side of Figure 6.1.[5] Skin regions supplied by sensory nerve fibers emanating at different levels of the spinal cord, called *dermatomes*, are shown in Figure 6.2.

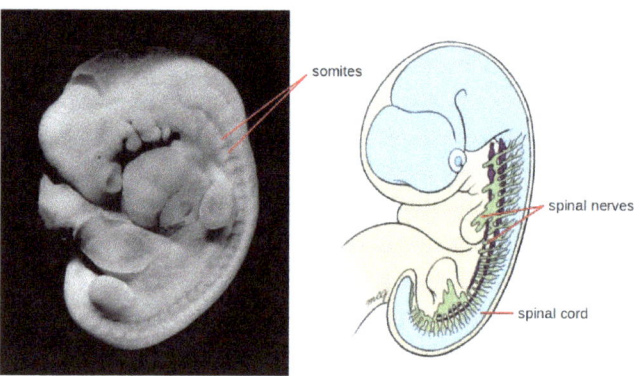

Figure 6.1: Segmentation in the human embryo (at five weeks)[6]

The general geometric pattern of dermatomes is that of transverse stripes, relative to the cranio-caudal axis of the young embryo. This pattern is maintained in the adult torso, and follows roughly along the adult limbs. However, the stripes along the limbs are not really straight but rather exhibit spiral forms.

This is because between the sixth and eighth weeks of embryo development the limbs rotate around their long axes, causing the originally-straight segmental pattern of lower limb innervation to twist into spirals.[8] This particular morphology is a consequence of functional adaptations from ancestral tetrapods, as shown in Figure 4.13 on page 42.

---

[4][Sch+15, p. 73]

[5][Sch+15, p. 238], [Kar15, p. 631]

[6]Adapted from *The Virtual Human Embryo* (www.ehd.org/virtual-human-embryo) and [Sch+15, fig. 4-22].

[7]Adapted from [Sch+15, fig. 20-28], [WPK09, fig. 4.7]. Also see [Pur+08, p. 209].

[8][Sch+15, p. 521]

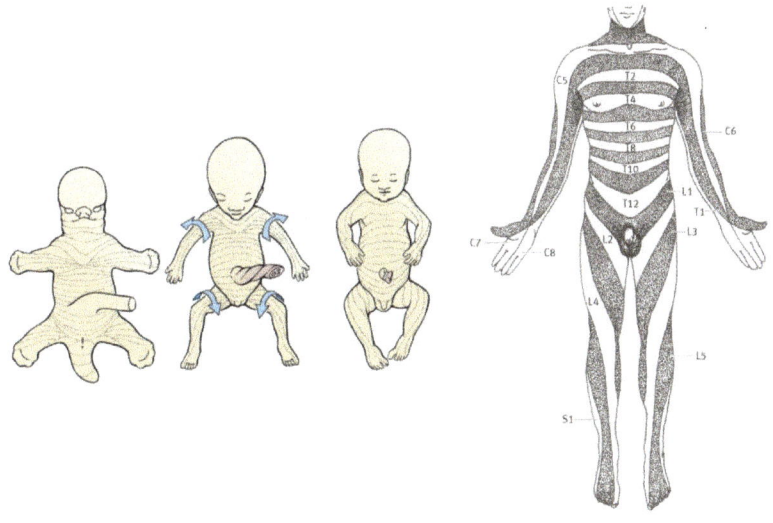

Figure 6.2: Geometric pattern of dermatomes[7]

6.2 Spinal Nerves

As discussed above, spinal nerves are bundles of nerve fibers that emanate laterally from the spinal cord (see top of Figure 6.3). They carry both motor and sensory fibers, which become merged into larger bundles by the time they emerge from the sides of the vertebral column (not shown in the figure).[9] The human body normally contains thirty-one spinal nerve pairs.[10]

The main trunk of a spinal nerve is relatively short, and starts dividing into branches shortly after it exits the vertebral column. Major branches that innervate the skin separate into dorsal, lateral and ventral sub-branches (rami), as shown by the different colors on the bottom left of Figure 6.3. This branching is the basis of a longitudinal division of the body surface into thirds, as shown on the bottom right of the figure.[11]

While dermatomes form a generally-horizontal pattern as seen in

---

[9][Kar15, p. 630], [Wan10, sec. 2.2]
[10][Wan10, secs. 2.2, 2.3]
[11][Wan10, sec. 2.3]

Figure 6.2 (given by the segmental arrangement of spinal nerves), there is also a subtle vertical pattern underlying it, as shown in Figure 6.3 (given by local branching of the spinal nerves). This anatomical aspect of the nervous system can be thought of as a conceptual grid.

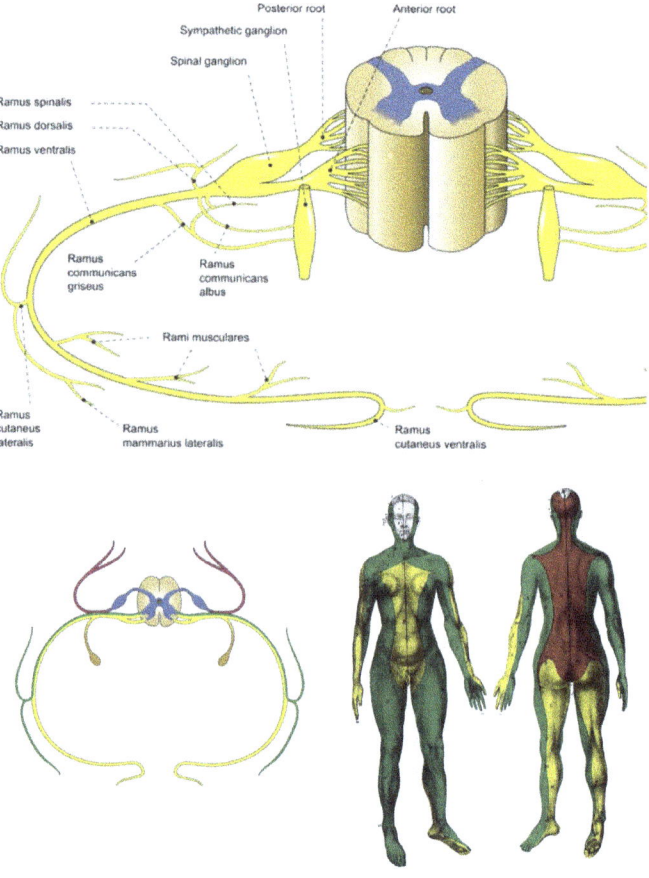

Figure 6.3: Configuration of spinal nerves[12]

---

[12]Adapted from [Wan10, figs. 2.4, 2.6a, 2.9].

6.2.1 Head Zones and Maximum Points

A schema that conforms conspicuously to dermatomes and the conceptual grid just described, is that of *Head zones*.[13] These are named after the previously-mentioned English neurologist Henry Head, who mapped areas of the skin that develop increased sensitivity and discomfort (allodynia) during the course of certain diseases.[14] In particular, he emphasized the existence of specific points of more pronounced sensitivity within these zones, that he called "maximum points" (see Figure 6.4).

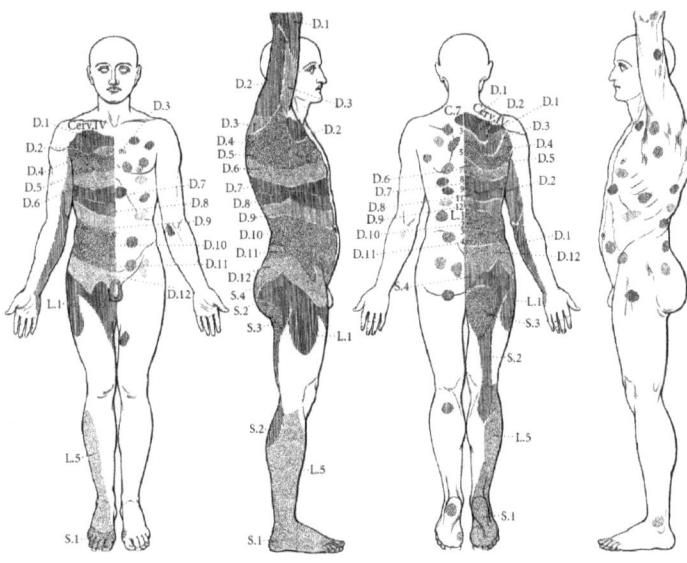

Figure 6.4: Head zones and maximum points[15]

In addition to producing detailed maps of zones and maximum points, Head also discovered (perhaps by accident) that external stimulation of maximum points sometimes relieved the patient's condition. He realized that he could do this via firm pressure in some cases, or by applying mustard ointment in others, to elicit the desired skin stimulation.

---

[13][Wan10, sec. 4.1]
[14][BHU11], [Wan10, secs. 4.1, 4.3]
[15]From [BHU11].

A study published in 2008 by Beissner et al.[16] analyzed Head's maximum points in patients suffering disease of specific organs (including lung, liver, stomach, kidney), and compared them against acupuncture points indicated in the treatment of the same organs. Remarkably, the study found very close correspondence between Head's maximum points and their homologous organ-specific acupuncture points.

These findings suggest that the viscerocutaneous reflexes described by Head are likely a rediscovery of what the Chinese had already observed and systematized more than two millennia earlier. As we will see next, in terms of salient anatomical and functional aspects, these reflexes seem to conform to the neural circuitry of the autonomic nervous system.[17]

6.3 Autonomic Nervous System

The *autonomic nervous system* is a subdivision of the nervous system (from a functional perspective) that regulates critical body functions such as blood pressure, heart rate, digestion, body temperature, metabolism and sexual arousal. It is a self-regulating control system that influences visceral (internal organ) operation via the activity of glands and smooth muscle fibers (found in the walls of hollow organs and blood vessels). For the most part, the autonomic system operates subconsciously and is not under voluntary control. The autonomic system comprises two major divisions: the sympathetic and parasympathetic subsystems.

When acting on glands, the autonomic system causes the release of a variety of neurotransmitters that affect multiple processes. This is a key function, given the challenge of maintaining homeostasis (the overall physiological balance of bodily functions) across the many organ systems of the body, in the presence of changing environmental conditions and dynamic behavior of people. For example, a mundane circumstance that requires an autonomic response is getting out of bed. Rising quickly from a lying position can produce a shift of roughly 0.5 liters of blood from the thorax and abdomen to the legs. This would entail an unsustainable drop in blood pressure, were it not for a compensatory constriction of blood vessels.[18] Even with the benefit of autonomic compensation, sometimes dizziness or fainting can be experienced.

---

[16][BHU11]
[17][Wan10, sec. 9.6]
[18][Pur+08, pp. 518, 536]

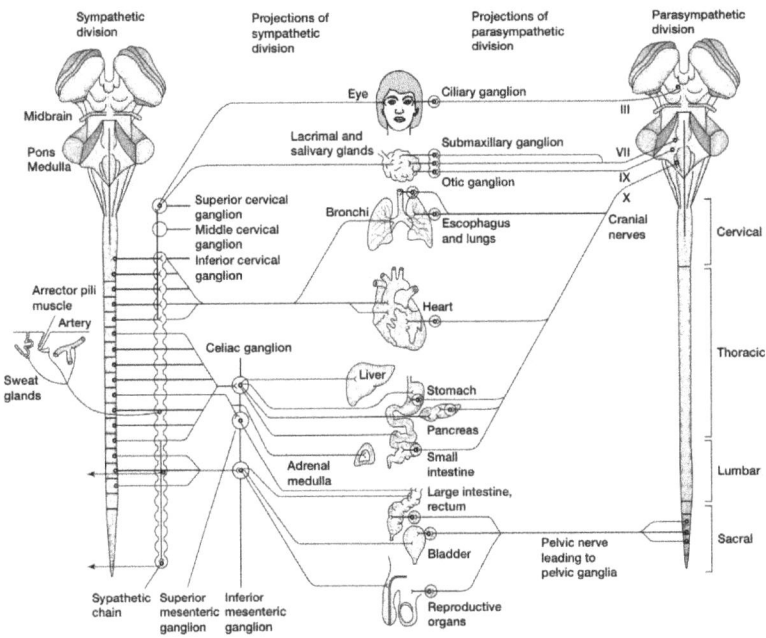

Figure 6.5: Circuitry of the autonomic nervous system[19]

Visceral anatomy does not follow a segmental pattern, and so innervation of the autonomic system would not be expected to follow a segmental pattern either.[20] However, visceral autonomic circuitry maps into the segmental distribution of spinal nerves quite evidently (on the sympathetic side) as shown in Figure 6.5.[21]

Generally speaking, organ sensory information induces local reflexes (at the level of the spinal cord) that modulate organ operations from moment to moment.[22] However, this control is not completely local. Higher integrative centers are involved in complex patterns of stimulation that require more comprehensive coordination, as seen in Figure 6.6.

---

[19]From [Kar15, fig. 16.21].
[20][Sch+15, p. 243], [Wan10, sec. 7.1]
[21][Kar15, p. 630]
[22][Pur+08, p. 524], [Kar15, pp. 650, 669]
[23]Adapted from [Pur+08, figs. 21.5, 21.7].

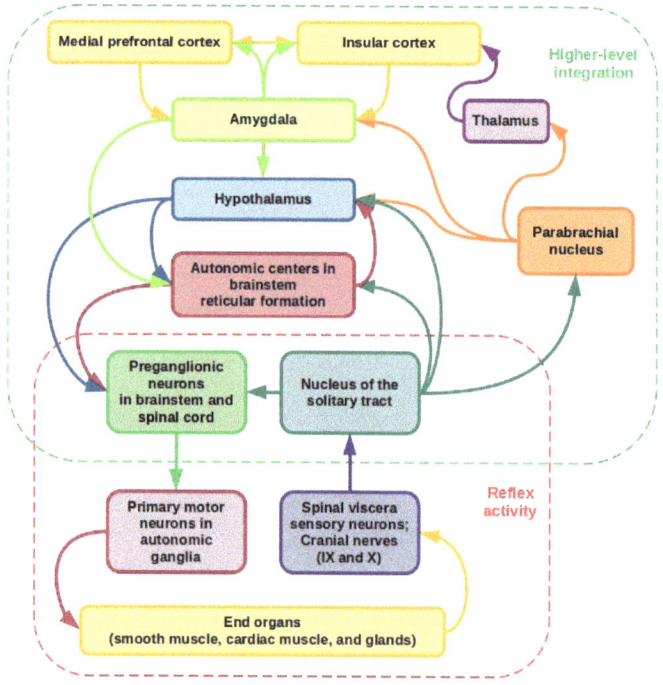

Figure 6.6: Autonomic network control[23]

Some functions of autonomic modules from Figure 6.6 are as follows:[24]

- **Nucleus of the solitary tract:** integrative center for reflexive control of visceral motor functions.
- **Parabrachial nucleus:** key relay of visceral sensory information to the amygdala, thalamus and hypothalamus.
- **Hypothalamus:** major locus of central control whose circuits receive sensory and contextual information, compare that information with biological set points, and activate relevant visceral motor, neuroendocrine, and somatic motor effector systems.
- **Medial prefrontal cortex and Insular cortex:** taken together these two cortical areas constitute a central autonomic network,

---

[24][Pur+08, pp. 513, 524–529], [MWK16, sec. 1.3]

which integrates visceral sensory and motor functions with information from cognitive centers that process semantic and emotional experiences.

The autonomic system is essentially organized around reflexes, which in engineering language can be understood as feedback control loops and system dynamics. Making use of systems thinking, we will see that some of these behaviors can be interpreted in terms of remarkably simple abstractions.

Chapter 7

Cybernetic Systems

Having brought up feedback control in the context of the autonomic nervous system, we dive into a set of associated disciplines in this chapter.

Feedback control is a circular regulation process, where sensed outputs of a system are transformed and fed back as inputs in order to maintain a targeted state.

Feedback mechanisms are widely used in modern technology for the stabilization of systems towards a certain goal, such as in satellite navigation, anti-aircraft fire control, and various robotic systems. In these conventional engineering applications, the disciplines that deal with the relevant modeling and mathematical methods are system dynamics and controls engineering.[1]

There are numerous biological phenomena which are consistent with the feedback arrangement. In a biology context the targeted state is called homeostasis, which as mentioned before corresponds to the set of conditions that allow an organism to remain alive, and, more importantly, healthy. In the case of complex animals, limits on these conditions are relatively narrow. For example, a variation of one-half degree Celsius in body temperature is generally considered a sign of illness, and a sustained variation of a few degrees can mean death. Similarly, hydrogen-ion concentration in the blood must be held within strict limits, leucocytes and other defenses against infection must be kept at adequate levels, and so on. Feedback regulation enables homeostatic control, and makes our

---

[1][Oga10]

body follow patterns that are characteristic of self-regulating systems.[2]

There is a discipline within systems theory called *cybernetics*, which specifically deals with feedback control mechanisms in animals. The cybernetic framework helps to characterize the regulatory structure of homeostatic mechanisms, even when detailed aspects remain unspecified and the system is mostly described in terms of input and output.[3]

In cybernetics, regulatory processes that maintain homeostasis are modeled by abstracting complex interactions of neuronal and biochemical networks into discrete functional components, which can be described concisely using powerful and versatile mathematical language (discussed in the following sections).

An important feature that biological systems share with engineered systems is that they are amenable to modular decomposition. This is used extensively in control and dynamic systems theory to make modeling of systems more tractable.

The typical first step in the modular decomposition of systems with feedback control is to isolate the process to be regulated, commonly referred to as the *plant*. The remaining components of the regulated system are then classified in terms of the functions they accomplish to facilitate the regulation.

Sensing and detection mechanisms constitute *sensing modules*, while mechanisms responsible for making decisions based on information provided by sensor modules constitute *controller modules*. The typical modular list of an engineering system also includes *actuation modules*. Actuation is necessary to transform the information-rich signal computed by the controllers into a quantity of sufficient magnitude to drive the plant in the desired direction.[4]

In this context, the structure presented in Figure 6.6 on page 66 can be interpreted as a feedback-control arrangement, whose bottom level is comprised of sensing and actuation modules. Superimposed on these are "reflex controllers", which mediate reflex-activity loops (outlined by the red-dashed box). These reflexes are in turn modulated by higher level centers (outlined by the green-dashed box), which can be referred to as "supervising controllers".[5]

It is plausible that acupuncture may be leveraging existing autonomic

---

[2][Wie65, sec. 4], [Ber69, p. 161]
[3][Wie65, sec. 1], [Ber69, p. 21]
[4][II10, p. 88]
[5][GK17]

control mechanisms to amplify self-repair capabilities of the body (we will get back to this in section 7.3). Using this idea to characterize acupuncture as a cybernetic mechanism, we can sketch out the following structure:

- surface anatomy innervation (in skin and superficial skeletal muscle) provides *sensory* functions

- visceral muscle and glands provide *actuation* functions (on organs, considered the target of actuation)

- peripheral nervous system, spinal cord and lower brainstem provide *reflex-control processing* functions

- brainstem, diencephalon and cerebral cortex provide *supervisory-control processing* functions

7.1 Nature of Feedback Processes

The essence of feedback processes is not very difficult to grasp, and can be adequately explained with an easy-to-visualize example.

There is a simple, well-known, purely mechanical feedback device called the *centrifugal governor*, which can be used to regulate the speed of a steam engine under varying conditions of load.[6] It consists of inertial blocks (masses) attached to pendulum rods that hang on the sides of a rotating shaft driven by the engine. The inertial blocks are drawn towards a low position due to their own weight (or constraining springs), but are swung upward by centrifugal action depending on the angular speed of the shaft. The blocks' changing position is transmitted by other rods to actuate the opening and closing of steam intake valves, which open when the engine slows down and the blocks fall, but close when the engine speeds up and the blocks rise. In other words, the engine reacts by speeding up when it experiences a slow-down, and by slowing down when it experiences a speed-up. So, in the presence of external disturbances, the engine manages to maintain a fairly constant speed. This type of action is known as *negative feedback*, and is the fundamental operational method used in controls engineering.

In the case of biological systems, the detailed mechanisms that support feedback processes might be different, but the basic system behavior can

---

[6][Wie65, sec. 4]

still be captured with the same mathematical abstractions (as we will see later on).

7.2 System Dynamics

To properly understand the dynamic behavior of a system, mathematical models are necessary. When building a model, abstractions are made to represent behavior of different elements within the system. Deciding what behaviors to emphasize (or even keep) in the model depends on what aspects of the system one is interested in.

For instance, consider a relatively complex machine such as an automobile. This machine can exhibit many different behaviors, but we could be interested in understanding how a passenger is bounced up and down when the vehicle drives over uneven terrain.

One important aspect of model building is to balance fidelity versus simplicity. The model could include a meticulous spatial (3D) description of inertial, compliant, and dissipative properties (among others) of a large number of components. Doing this is pertinent under some circumstances, such as when experts who already understand the intricacies of very similar machines perform detailed analysis. But if the study is exploratory, too many variables and details might obscure key underlying behaviors. Figuring out what to focus on, and choosing the proper level of fidelity, is very much an art.

But let's say that after meditating on the matter, we have concluded that the appropriate degree of complexity for our exploration is a model that studies vertical displacement of the car only, and that one-quarter of the car is representative enough of the entire four-wheeled vehicle. Then, we lump properties of different components into a handful of system elements endowed with discrete quantifiable parameters, and construct what is known as a *lumped parameter model*. The result is shown on the left side of Figure 7.1.

It is possible to obtain appropriate mathematical equations that characterize the system directly from this type of model sketch, but additional insight can be gained by translating the description into a *bond graph* (such as the one shown on the right side of the figure). Bond graphs are important in systems theory because their language of abstraction is quite universal. Using the language of bond graphs, the representation of system elements is made uniform across different domains, and this

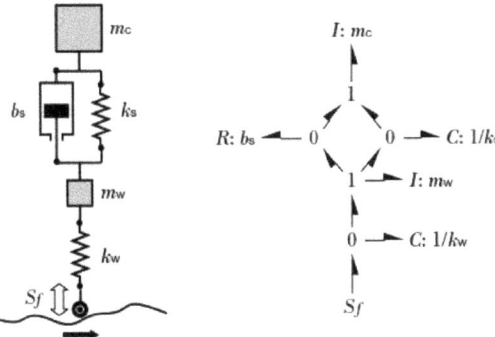

Figure 7.1: Lumped parameter model and bond graph of an automobile suspension system

facilitates recognition of patterns and homologies between systems that could be very different in outer appearance.

Bond graphs are not only used in conventional engineering problems, but have also been leveraged to study systems ranging from financial structures to hierarchies of biochemical networks.[7]

We are using a mechanical example here because most people will likely find it easier to visualize than, say, a biochemical process. However, we will remap the formulations onto the biological domain shortly.

The bond graph in Figure 7.1 contains elements that behave inertially (frame mass m_c and wheel mass m_w), capacitatively (suspension stiffness k_s and tire stiffness k_w), and dissipatively (suspension damping b_s). The input to this system is vertical motion imposed by the irregular geometry of the ground (identified as a flow source S_f). The elements are connected with 0 or 1 *junctions*, which indicate whether effort (force) or flow (velocity) is conserved through that junction.

With the bond graph in place, it is fairly straightforward to write down the mathematical formulation that dictates how the system behaves. In fact, there are several software programs available that do this directly from the bond graph.[8]

This kind of model produces a set of differential equations, which

---

[7][GCC15], [DP11]
[8][DP11]

constrains the time evolution of what are called *state variables*. These are key quantities that characterize the system's state. The choice of state variables for any given system is not unique, but a practical selection for the above system is as follows:

- x_1: frame displacement
- x_2: frame velocity
- x_3: wheel displacement
- x_4: wheel velocity

An important reason for this choice of state variables is that they allow the system's dynamics to be expressed in terms of first-order coupled differential equations of the form:

$$
\begin{aligned}
\dot{x}_1(t) &= f_1(x_1, x_2, \ldots, x_n; \ u_1, u_2, \ldots, u_r; \ t) \\
\dot{x}_2(t) &= f_2(x_1, x_2, \ldots, x_n; \ u_1, u_2, \ldots, u_r; \ t) \\
&\vdots \\
\dot{x}_n(t) &= f_n(x_1, x_2, \ldots, x_n; \ u_1, u_2, \ldots, u_r; \ t)
\end{aligned}
\tag{7.1}
$$

where $\dot{x}_i$ are time derivatives of the state variables, u_i are inputs to the system, t is time and f_i are functions that depend on x_i, u_i and t. Additionally, for the sake of generality and convenience, a number of output variables y_i is also defined, with dependence on x_i, u_i and t. Using vector notation, the overall dynamic description of the system, or *state-space equations*, can be written as:

$$
\dot{\mathbf{x}}(t) = \mathbf{f}(\mathbf{x}, \mathbf{u}, t) \tag{7.2a}
$$

$$
\mathbf{y}(t) = \mathbf{g}(\mathbf{x}, \mathbf{u}, t) \tag{7.2b}
$$

Formally, 7.2a is called the state equation and 7.2b is the output equation.[9] This *nonlinear* form of the state-space equations is very comprehensive, but can sometimes also be unnecessarily difficult to digest (computationally and intuitively). It is common practice in engineering to *linearize* the equations in the vicinity of an operational point, thus simplifying the expressions substantially.[10] In addition, the behavior

---

[9][Oga10, pp. 30-31]
[10][Oga10, p. 43], [KMR12, p. 167]

of linearized equations can provide valuable information and insights regarding the system being modeled.[11]

Having said that, in our example we don't need to perform any linearization because the lumped parameter model is already linear. This means that the above equations can be directly expressed in matrix form, as *linear state-space equations*:

$$\dot{\mathbf{x}} = \mathbf{A}\mathbf{x} + \mathbf{B}\mathbf{u}$$
$$\mathbf{y} = \mathbf{C}\mathbf{x} + \mathbf{D}\mathbf{u} \tag{7.3}$$

We can obtain an instantiation of Equation 7.3 for the model shown in Figure 7.1 by using the state variables defined earlier. As pointed out above, the system input is just a single variable (which we call u_1) that expresses the vertical displacement imposed by terrain geometry. Also, to keep things simple we make the output vector $\mathbf{y}$ identical to the state vector $\mathbf{x}$, so matrix $\mathbf{C}$ is the identity matrix and matrix $\mathbf{D}$ is zero. Then, the behavior of the lumped parameter model is dictated by:

$$\begin{bmatrix} \dot{x}_1 \\ \dot{x}_2 \\ \dot{x}_3 \\ \dot{x}_4 \end{bmatrix} = \begin{bmatrix} 0 & 1 & 0 & 0 \\ -\dfrac{k_s}{m_c} & -\dfrac{b_s}{m_c} & \dfrac{k_s}{m_c} & \dfrac{b_s}{m_c} \\ 0 & 0 & 0 & 1 \\ \dfrac{k_s}{m_w} & \dfrac{b_s}{m_w} & -\dfrac{k_s+k_w}{m_w} & -\dfrac{b_s}{m_w} \end{bmatrix} \begin{bmatrix} x_1 \\ x_2 \\ x_3 \\ x_4 \end{bmatrix} + \begin{bmatrix} 0 \\ 0 \\ 0 \\ -\dfrac{k_w}{m_w} \end{bmatrix} \begin{bmatrix} u_1 \end{bmatrix}$$

$$\begin{bmatrix} y_1 \\ y_2 \\ y_3 \\ y_4 \end{bmatrix} = \begin{bmatrix} 1 & 0 & 0 & 0 \\ 0 & 1 & 0 & 0 \\ 0 & 0 & 1 & 0 \\ 0 & 0 & 0 & 1 \end{bmatrix} \begin{bmatrix} x_1 \\ x_2 \\ x_3 \\ x_4 \end{bmatrix} + \begin{bmatrix} 0 \\ 0 \\ 0 \\ 0 \end{bmatrix} \begin{bmatrix} u_1 \end{bmatrix} \tag{7.4}$$

Notice that matrix $\mathbf{A}$ (called the state matrix) has several non-zero entries outside of its main diagonal (which runs from top left to bottom right of the matrix). This characterizes the *coupling* between state variables. The presence of coupling means that as the value of one of the x_i changes, it affects the values of the others.

---

[11][II10, p. 10]

To illustrate this kind of behavior, we can run a short computer simulation of Equation 7.4 using some typical values for parameters m_c, m_w, k_s, k_w, b_s. Figure 7.2 shows a simulation where the state variables interact in the absence of any system input u_1, starting from arbitrary initial conditions.

As you can see from the result, even for this relatively simple model it would be rather difficult to predict precisely how the state variables interact and combine, without the aid of a computer. This is where computers become invaluable: they enable us to explore and gain insight into system behavior quickly and easily. Mathematical models and computers greatly enhance our basic capabilities of understanding and intuition.

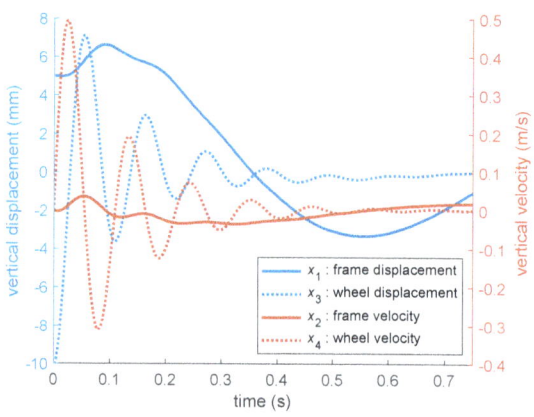

Figure 7.2: Simulation of automobile suspension system

7.3 Control in Biological Systems

As mentioned earlier, some biological arrangements like the autonomic system exhibit self-regulatory behavior, where the system acts in a way to maintain one or more variables near desired values. In that kind of situation it is customary to depict the self-regulatory system behavior using *feedback loops*, such as the one shown in Figure 7.3.

State-space equations can be derived from this kind of diagram, or via a bond graph. Either way, in order to perform system simulations, mathematical expressions like state-space equations (or equivalent) are necessary.

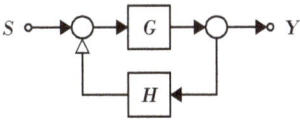

Figure 7.3: Closed-loop feedback model

In closed-loop schemes like that represented by Figure 7.3, the controlling input S is a "setting" that may change over time. For example, let's assume that a quantity controlled by the autonomic system, such as the heart rate, can be modeled by this type of scheme. In the course of daily life the (homeostatic) setting could change drastically, depending on the situation. For instance, in transitioning from a slow walk to a fast run, the heart rate would need to rise significantly from a low value to a high value (due to an increased demand of oxygen to support muscle metabolism, among other things). The way in which the heart rate settles onto a new value will depend on various factors, such as size and fiber composition of leg muscles, fitness of heart and lungs, blood biochemistry, etc. In some cases the heart rate will monotonically increase towards its new steady state, while in other cases it will overshoot and then fluctuate somewhat before stabilizing.

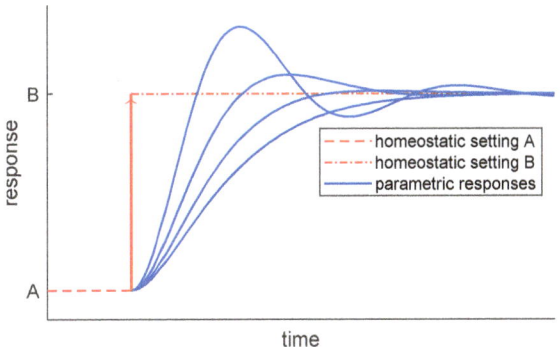

Figure 7.4: Simulation of parametric step responses

Figure 7.4 shows a range of such potential behaviors, based on a simple lumped parameter model (similar to the earlier suspension model). In this model, the different responses arise when internal parameters (such as the inertial, capacitive and dissipative elements that were part of the suspension model) are modified.

There may also be a different kind of scenario, where the targeted homeostatic setting remains constant but a temporary event perturbs the system. For example, as mentioned in section 6.3, standing up quickly from a lying position causes a large sudden change in blood flow, but blood pressure needs to be maintained as constant as possible (to avoid dizziness or fainting). In such a case it is convenient to depict the feedback mechanism as in Figure 7.5, where the control input S is a constant homeostatic setting and the standing-up event is considered a disturbance D.

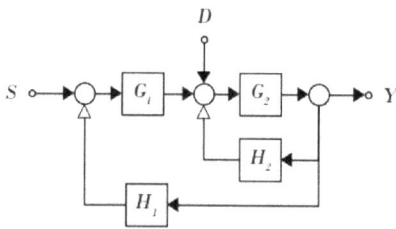

Figure 7.5: Closed-loop feedback model with disturbance

Like before, relevant mathematical models can be developed and simulated. Figure 7.6 shows the response of such as system for a couple of different instantiations of system parameters. In this case, the system equations used to generate the curves actually correspond to a simplified version of the mechanical model from Figure 7.1 on page 73 (using the top half of the suspension model, to be specific).[12]

In the context of an autonomic effect, the green curve could be interpreted as a reference response. Depending on how the system is parameterized, the modification of one or more system parameter values could result in a substantially different response. This is illustrated by the blue curve, whose maximum amplitude is almost 2.5 times the value of the reference curve's amplitude.

---

[12]A simple mass-spring-damper model.

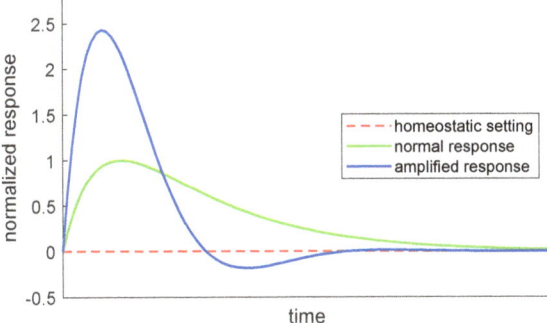

Figure 7.6: Disturbance-response simulation with a reduced model

7.3.1 Hemodynamic Response Function

Cerebral blood flow (CBF) has been a topic of study in neurophysiology for decades. One well-known characteristic of cerebral blood flow is that it is affected by the intake of commonly used substances (e.g., caffeine, nicotine, and alcohol), by changes in the concentration of endogenous substances (e.g., estrogen and adrenaline), administration of various drugs (e.g., cocaine or acetazolamide), and by the use of anesthetic agents.[13]

Even though the precise mechanisms that drive these hemodynamic effects have not been elucidated, some clinical aspects have been clearly characterized experimentally. For example, different levels of CBF responses have been correlated to different concentrations of carbon dioxide in the blood.

In experiments performed by Cohen et al.[14] at the University of Minnesota, volunteers were "primed" before their CBFs were tested, by inducing different levels of carbon dioxide in their blood. Specifically, subjects were tested under *normocapnia* (normal carbon dioxide levels), *hypocapnia* (low carbon dioxide levels induced by hyperventilation) and *hypercapnia* (high carbon dioxide levels induced by inhalation of air infused with 5% CO_2).

Under each one of those conditions, the volunteers' neural activity was perturbed with a consistent set of visual stimuli. The stimuli elicited increased activity in the visual cortex, as would normally be expected, but

---

[13][CUK02]
[14][CUK02]

with distinct dynamic characteristics that depended on the pre-induced carbon dioxide level. This is shown on the left side of Figure 7.7.

Even though the causal circumstances underlying the relevant neural activities are still a matter of debate,[15] it is clear that the magnitude and dynamic characteristics of the response are a function of carbon dioxide concentration in the blood. Notably, the basic behavior of this hemodynamic response is isomorphic to that of the simple lumped parameter model underlying Figure 7.6.

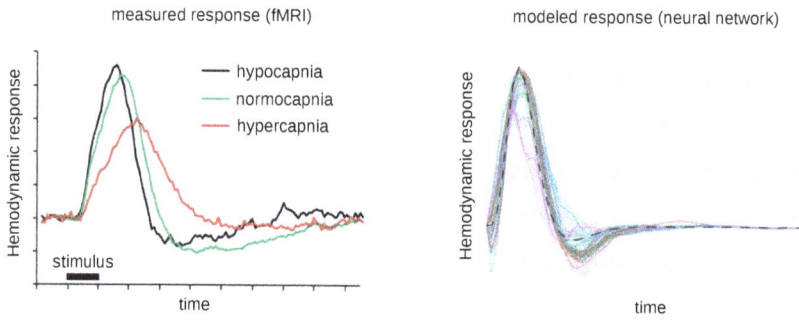

Figure 7.7: Hemodynamic response[16]

This type of homeostatic event has been the subject of research ever since fMRI was developed in 1990 by Seiji Ogawa at Bell Labs.[17] The general shape of the response, called *hemodynamic response function* (HRF), has been emulated in various ways, including analytic gamma functions and deep neural networks.[18] An example of HRF emulations using neural networks is shown on the right side of Figure 7.7.

While the overshoots, undershoots and time dilations of the observed hemodynamic responses can be captured using transformations such as gamma functions or artificial neural networks, they can also be directly associated with inertial, capacitive and dissipative behavior in models that are conceptually much simpler (as seen in Figure 7.6). Importantly, the

---

[15][CUK02], [KO12]

[16]Adapted from [KO12] and [GG17].

[17]fMRI (functional magnetic resonance imaging) is a technique based on MRI, capable of measuring changes associated with cerebral blood flow.

[18][Lin+09], [GG17]

more complex models are perhaps not particularly insightful in a causal sense, whereas properly reduced models may bring essential underlying elements to the foreground and can provide some degree of intuition. It is commonly the case that simplicity of abstraction facilitates breakthroughs that lead to better understanding.

7.3.2 Interpretations

Dynamic behavior such as that shown in Figure 7.6 could be used to describe different types of analogous biological responses, like the concentration of immune cells during the course of a viral infection.[19]

And perhaps the mechanism by which acupuncture operates can be explained by a similar system effect. Let us suppose that baseline autonomic response to some visceral condition can be captured by the green curve in Figure 7.6. Stimulation from acupuncture may be influencing high-level integration centers (like those shown in Figure 6.6 on page 66) in a way that modifies their supervisory functions over the autonomic system. This could result in a modification of "system parameters", that in turn manifests as an amplification of latent restorative functions, resulting in something like the blue curve.

So conceivably, the autonomic system could be "preadapted" to the stimulations that acupuncture provides. In that sense, acupuncture can be seen as a kind of *hack* of the human body, that manages to elicit some of its self-repair capabilities.

In terms of latent self-repair potential, consider that some animals can regrow their tails, and even entire limbs. Remarkably, a similar capability has been observed in humans, although to a more limited extent (this kind of spontaneous regeneration is only seen in children's fingertips, and as long as the proximal nail matrix remains intact).[20] This is not to say that acupuncture evokes this particular kind of mechanism. But it suggests that the arsenal of restorative capabilities of complex organisms might be much deeper than we think.

Having hinted at the hypothesis of a magnified self-repair mechanism homologous to modulated hemodynamic reflexes, there is also the possibility that acupuncture reflects multiple, relatively independent mechanisms.

---

[19][Su+09]
[20][Dav19, pp. 114-115], [Keo14, pp. 1-3]

As a relevant point, we should note that the autonomic system is actually conformed by a network of self-regulating mechanisms. For example, when we are a bit chilly, changes in brain hormones lead to constriction of blood vessels. This constriction decreases blood flow and raises blood pressure in the skin, which in turn enhances the heat insulation of superficial tissues. But if that warming method is not sufficient, we start producing heat by shivering, which is an entirely different response mechanism. These various mechanisms are not completely independent but couple to each other to form nested and interlocking information flows.[21]

Finally, we should also mention that the curative effects of therapies such as acupuncture are not just limited to organs that perform the more immediate physical functions, but can be targeted to mental states as well. For example, acupuncture has been effectively used in the treatment of depression.[22]

7.4   Reverse Engineering

Biology, at its core, is a reverse engineering endeavor. Unlike traditional engineering disciplines where the emphasis is on design (forward engineering), biology takes the existing designs of living organisms and seeks to deduce their working principles.

Disciplines like cybernetics provide powerful homologies between living organisms and well-developed engineering fields, and the resulting parallels and analogies could be instrumental in unlocking challenging and enigmatic biological mechanisms.

We have seen that even simple models can provide important insights. In addition, the computational capabilities available to us in this *second machine age* give us the ability to study more complex system behaviors. Techniques such as bond graphs and state-space models can be easily extended to much more complex systems than the ones presented here. In the context of systems thinking, tools like these can be quite powerful. Unexpected processes may become evident under the right lens.

Having said that, it may be quite challenging in some cases to reverse-engineer the underlying rules that make a system tick. For instance, consider the patterns in Wolfram's cellular automaton from Figure 4.6

---

[21][Voi20, ch. 2], [Dav19, p. 90]
[22][Org03]

on page 29. It is not very difficult to imagine a mechanism capable of leveraging the observed repeating patterns for some systemic effect, but it would be exceedingly troublesome to try decoding the discrete rules that enabled the emergence of those patterns.

From that perspective, systems theory can also be useful in a practical sense by enabling "explanation in principle" of complex phenomena.[23] Of course, ultimately a quantifiable model needs to be developed in order to claim true understanding of a mechanism, but systems thinking is a powerful way to establish viable qualitative models. The right *grey box* models facilitate the crystallizing of *white box* models.

---

[23][Ber69, pp. 36, 113]

Epilogue

Although biological entities and engineered technologies have very different detailed implementations, they can be quite similar in their systems-level organization. When viewed under the lens of systems theory, living organisms share some important structural features with engineered systems, such as modularity, robustness and use of recurring circuit elements.[24]

It seems therefore natural that systems engineering and related disciplines can play a major role in the understanding of biological systems. In particular, since feedback regulation is an essential function in living organisms, the use of engineering tools from areas such as system dynamics and controls could prove invaluable in the study of such systems.

As this short survey shows, an extraordinary amount of knowledge has been accumulated in various fields and disciplines. However, our understanding of nature and reality is still relatively limited, and we must accept the fact that there is much that we don't know. In this regard, Plato's *allegory of the cave* reminds us of the innate limitations of human perception: our conception of reality is shaped by shadows of a world that lies beyond our senses.[25]

Nonetheless, it is in our nature to keep pushing the rock up the mountain of knowledge. This allows us to undertake the most exquisite and noble of tasks: to transmit our experience and insight to those who follow.[26]

And of course, much fun is to be had in learning and reveling in the wonders of the world!

---

[24][VB13]
[25][Pla68, book VII]
[26][Gre07, p. 22]

www.ingramcontent.com/pod-product-compliance
Lightning Source LLC
Chambersburg PA
CBHW040947170526
45162CB00001B/1